MATH 100
Teacher's Guide Part 1

Author:

Carol Bauler, B.A.

Editor:

Alan Christopherson, M.S.

Media Credits:

Pages 26, 68, 108, 150, 190: © wenchiawang, iStock, Thinkstock; **87:** © blueringmedia, iStock, Thinkstock; **125:** © Gurzzza, iStock, Thinkstock; © Elvetica, iStock, Thinkstock; © Askold Romanov, Hemera, Thinkstock; PrettyVectors, iStock, Thinkstock; **138:** © Allevinatis, iStock, Thinkstock; © mocoo, iStock, Thinkstock; **140, 142, 153:** © Kazberry, iStock, Thinkstock; **177:** © RobinOlimb, iStock, Thinkstock; **177, 209:** © Volha Hlinskaya, iStock, Thinkstock; **218:** © Ashva73, iStock, Thinkstock.

Alpha Omega

PUBLICATIONS

804 N. 2nd Ave. E.
Rock Rapids, IA 51246-1759

© MCMXCVII by Alpha Omega Publications, Inc. All rights reserved.
LIFEPAC is a registered trademark of Alpha Omega Publications, Inc.

MATH 100

LIFEPAC® Overview

MATH SCOPE & SEQUENCE

KINDERGARTEN

Lessons 1–40	Lessons 41–80	Lessons 81–120	Lessons 121–160
Directions – right, left, high, low, etc.	**Directions** – right, left, high, low, etc.	**Directions** – right, left, high, low, etc.	**Directions** – right, left, high, low, etc.
Comparisons – big, little, alike, different	**Comparisons** – big, little, alike, different	**Comparisons** – big, little, alike, different	**Comparisons** – big, little, alike, different
Matching	**Matching**	**Matching**	**Matching**
Cardinal Numbers – to 9	**Cardinal Numbers** – to 12	**Cardinal Numbers** – to 19	**Cardinal Numbers** – to 100
Colors – red, blue, green, yellow, brown, purple	**Colors** – orange	**Colors** – black, white	**Colors** – pink
Shapes – circle, square, rectangle, triangle	**Shapes** – circle, square, rectangle, triangle	**Shapes** – circle, square, rectangle, triangle	**Shapes** – circle, square, rectangle, triangle
Number Order	**Number Order**	**Number Order**	**Number Order**
Before and After	**Before and After**	**Before and After**	**Before and After**
Ordinal Numbers – to 9th	**Ordinal Numbers** – to 9th	**Ordinal Numbers** – to 9th	**Ordinal Numbers** – to 9th
Problem Solving	**Problem Solving**	**Problem Solving**	**Problem Solving**
	Number Words – to nine	**Number Words** – to nine	**Number Words** – to nine
	Addition – to 9	**Addition** – multiples of 10	**Addition** – to 10 and multiples of 10
		Subtraction – to 9	**Subtraction** – to 10
		Place Value	**Place Value**
		Time/Calendar	**Time/Calendar**
			Money
			Skip counting – 2's, 5's, 10's
			Greater/Less Than

MATH SCOPE & SEQUENCE

	Grade 1	Grade 2	Grade 3
UNIT 1	**NUMBER ORDER, ADD/SUBTRACT** • Number order, skip count • Add, subtract to 9 • Story problems • Measurements • Shapes	**NUMBERS AND WORDS TO 100** • Numbers and words to 100 • Operation symbols: +, –, =, >, < • Add and subtract • Place value and fact families • Story problems	**ADD/SUB TO 18 AND PLACE VALUE** • Digits, place value to 999 • Add and subtract • Linear measurements • Operation symbols: +, –, =, ≠, >, < • Time
UNIT 2	**ADD/SUBTRACT TO 10, SHAPES** • Add, subtract to 10 • Number words • Place value • Patterns, sequencing, estimation • Shapes	**ADD/SUBTRACT AND EVEN/ODD** • Numbers and words to 200 • Add, subtract, even and odd • Skip count 2, 5, and 10 • Ordinal numbers, fractions, and money • Shapes	**CARRYING AND BORROWING** • Fact families, patterns, and fractions • Add and subtract with carrying and borrowing • Skip count 2, 5, and 10 • Money, shapes, lines • Even and odd
UNIT 3	**FRACTIONS, TIME, AND SYMBOLS** • Number sentences • Fractions • Story problems • Time and the = symbol • Oral directions	**ADD WITH CARRYING TO THE 10'S PLACE** • Add with carrying to the 10's place • Subtract • Flat shapes, money, A.M./P.M. • Rounding to the 10's place • Standard measurements	**FACTS OF ADD/SUB AND FRACTIONS** • Add 3 numbers w/ carrying • Coins, weight, volume, A.M./P.M. • Fractions • Skip count 3, subtract w/ borrowing • Oral instructions
UNIT 4	**ADD TO 18, MONEY, MEASUREMENT** • Add to 18 • Skip count, even and odd • Money • Shapes and measurement • Place value	**NUMBERS/WORDS TO 999, AND GRAPHS** • Numbers and words to 999 • Addition, subtraction, and place value • Calendar • Measurements and solid shapes • Making change	**ROUND, ESTIMATE, STORY PROBLEMS** • Place value to 9,999 • Rounding to the 10's and estimating • Add and subtract fractions • Roman numerals • 1/4 inch
UNIT 5	**COLUMN ADDITION AND ESTIMATION** • Add three 1-digit numbers • Ordinal numbers • Time and number lines • Estimation and charts • Fractions	**ADD/SUBTRACT TO THE 100'S PLACE** • Data and bar graphs and shapes • Add and subtract to the 100's • Skip count 3 and place value to the 100's • Add fractions • Temperature	**PLANE SHAPES AND SYMMETRY** • Number sentences • Rounding to the 100's and estimation • Perimeter and square inch • Bar graph, symmetry, and even/odd rules • Temperature
UNIT 6	**NUMBER WORDS TO 99** • Number words to 99 • Add two 2-digit numbers • Symbols: > and < • Fractions • Shapes	**SUBTRACT WITH BORROWING FROM 10'S** • Measurements • Time and money • Subtract w/ borrowing from the 10's place • Add and subtract fractions • Perimeter	**MULTIPLICATION, LINES, AND ANGLES** • Add and subtract to 9,999 • Multiples and multiplication facts for 2 • Area and equivalent fractions • Line graphs, segments, and angles • Money
UNIT 7	**COUNT TO 200, SUBTRACT TO 12** • Number order and place value • Subtract to 12 • Operation signs • Estimation and time • Graphs	**ADD WITH CARRYING TO THE 100'S PLACE** • Add with carrying to the 100's place • Fractions as words • Number order in books • Rounding and estimation	**ADD/SUB MIXED NUMBERS, PROBABILITY** • Multiplication facts for 5 and missing numbers • Add and subtract mixed numbers • Subtract with 0s in the minuend • Circle graphs • Probability
UNIT 8	**ADD/SUBTRACT TO 18** • Addition, subtract to 18 • Group counting • Fractions • Time and measurements • Shapes	**VOLUME AND COIN CONVERSION** • Addition, subtraction, and measurements • Group counting and "thinking" answers • Convert coins • Directions – North, South, East, and West • Length and width	**MEASUREMENTS AND MULTIPLICATION** • Multiplication facts for 3 & 10, multiples of 4 • Convert units of measurement • Decimals and directions • Picture graphs and missing addends • Length and width
UNIT 9	**SENSIBLE ANSWERS** • Fact families • Sensible answers • Subtract 2-digit numbers • Add three 2-digit numbers	**AREA AND SQUARE MEASUREMENT** • Area and square measurement • Add three 2-digit numbers with carrying • Add coins and convert to cents • Fractions and quarter-inches	**MULT, METRICS, AND PERIMETER** • Add and subtract whole numbers, fractions, and mixed numbers • Standard measurements and metrics • Operation symbols • Multiplication facts for 4
UNIT 10	**REVIEW** • Addition, subtraction, and place value • Directions – North, South, East, and West • Fractions • Patterns	**REVIEW** • Rules for even and odd numbers • Round numbers to the 100's place • Digital clocks and sensible answers • Add three 3-digit numbers	**PROBABILITY, UNITS, AND SHAPES** • Addition and subtraction • Rounding to the 1,000's place and estimating • Probability, equations, and parentheses • Perimeter and area • Multiplication facts for 2, 3, 4, 5, and 10

MATH SCOPE & SEQUENCE

Grade 4	Grade 5	Grade 6	
WHOLE NUMBERS AND FRACTIONS • Naming whole numbers • Naming fractions • Sequencing patterns • Numbers to 1,000	**PLACE VALUE, ADDITION, AND SUBTRACTION** • Place value • Rounding and estimating • Addition • Subtraction	**WHOLE NUMBERS AND ALGEBRA** • Whole numbers and their properties • Operations and number patterns • Algebra	UNIT 1
MULTIPLYING WHOLE NUMBERS • Operation symbols • Multiplication — 1-digit multipliers • Addition and subtraction of fractions • Numbers to 10,000	**MULTIPLYING WHOLE NUMBERS AND DECIMALS** • Multiplying whole numbers • Powers • Multiplying decimals	**DATA ANALYSIS** • Collecting and describing data • Organizing data • Displaying and interpreting data	UNIT 2
SEQUENCING AND ROUNDING • Multiplication with carrying • Rounding and estimation • Sequencing fractions • Numbers to 100,000	**DIVIDING WHOLE NUMBERS AND DECIMALS** • One-digit divisors • Two-digit divisors • Decimal division	**DECIMALS** • Decimal numbers • Multiplying and dividing decimal numbers • The metric system	UNIT 3
LINES AND SHAPES • Plane and solid shapes • Lines and line segments • Addition and subtraction • Multiplication with carrying	**ALGEBRA AND GRAPHING** • Expressions • Functions • Equations • Graphing	**FRACTIONS** • Factors and fractions • The LCM and fractions • Decimals and fractions	UNIT 4
DIVISION AND MEASUREMENTS • Division – 1-digit divisor • Families of facts • Standard measurements • Number grouping	**MEASUREMENT** • The metric system • The customary system • Time • Temperature	**FRACTION OPERATIONS** • Adding and subtracting fractions • Multiplying and dividing fractions • The customary system	UNIT 5
DIVISION, FACTORS, AND FRACTIONS • Division – 1-digit divisors with remainders • Factors and multiples • Improper and mixed fractions • Equivalent fractions	**FACTORS AND FRACTIONS** • Factors • Equivalent fractions • Fractions	**RATIO, PROPORTION, AND PERCENT** • Ratios • Proportions • Percent	UNIT 6
WHOLE NUMBERS AND FRACTIONS • Multiplication – 2-digit multipliers • Simplifying fractions • Averages • Decimals in money problems • Equations	**FRACTION OPERATIONS** • Like denominators • Unlike denominators • Multiplying fractions • Dividing fractions	**PROBABILITY AND GEOMETRY** • Probability • Geometry: Angles • Geometry: Polygons	UNIT 7
WHOLE NUMBERS AND FRACTIONS • Division – 1-digit divisors • Fractions and unlike denominators • Metric units • Whole numbers: +, −, ×, ÷	**DATA ANALYSIS AND PROBABILITY** • Collecting data • Analyzing data • Displaying data • Probability	**GEOMETRY AND MEASUREMENT** • Plane figures • Solid figures	UNIT 8
DECIMALS AND FRACTIONS • Reading and writing decimals • Adding and subtracting mixed numbers • Cross multiplication • Estimation	**GEOMETRY** • Geometry • Classifying plane figures • Classifying solid figures • Transformations • Symmetry	**INTEGERS AND TRANSFORMATIONS** • Integers • Integer operations • Transformations	UNIT 9
ESTIMATION, CHARTS, AND GRAPHS • Estimation and data gathering • Charts and graphs • Review numbers to 100,000 • Whole numbers: +, −, ×, ÷	**PERIMETER, AREA, AND VOLUME** • Perimeter • Area • Surface area • Volume	**EQUATIONS AND FUNCTIONS** • Equations • More equations and inequalities • Functions	UNIT 10

MATH SCOPE & SEQUENCE

	Grade 7	Grade 8	Grade 9
UNIT 1	**INTEGERS** • Adding and Subtracting Integers • Multiplying and Dividing Integers • The Real Number System	**THE REAL NUMBER SYSTEM** • Relationships • Other Forms • Simplifying	**VARIABLES AND NUMBERS** • Variables • Distributive Property • Definition of signed numbers • Signed number operations
UNIT 2	**FRACTIONS** • Working with Fractions • Adding and Subtracting Fractions • Multiplying and Dividing Fractions	**MODELING PROBLEMS IN INTEGERS** • Equations with Real Numbers • Functions • Integers • Modeling with Integers	**SOLVING EQUATIONS** • Sentences and formulas • Properties • Solving equations • Solving inequalities
UNIT 3	**DECIMALS** • Decimals and Their Operations • Applying Decimals • Scientific Notation • The Metric System	**MODELING PROBLEMS WITH RATIONAL NUMBERS** • Number Theory • Solving Problems with Rational Numbers • Solving Equations and Inequalities	**PROBLEM ANALYSIS AND SOLUTION** • Words and symbols • Simple verbal problems • Medium verbal problems • Challenging verbal problems
UNIT 4	**PATTERNS AND EQUATIONS** • Variable Expressions • Patterns and Functions • Solving Equations • Equations and Inequalities	**PROPORTIONAL REASONING** • Proportions • Percents • Measurement/Similar Figures	**POLYNOMIALS** • Addition of polynomials • Subtraction of polynomials • Multiplication of polynomials • Division of polynomials
UNIT 5	**RATIOS AND PROPORTIONS** • Ratios, Rates, and Proportions • Using Proportions • Fractions, Decimals, and Percents	**MORE WITH FUNCTIONS** • Solving Equations • Families of Functions • Patterns	**ALGEBRAIC FACTORS** • Greatest common factor • Binomial factors • Complete factorization • Word problems
UNIT 6	**PROBABILITY AND GRAPHING** • Probability • Functions • Graphing Linear Equations • Direct Variation	**MEASUREMENT** • Angle Measures and Circles • Polygons • Indirect Measure	**ALGEBRAIC FRACTIONS** • Operations with fractions • Solving equations • Solving inequalities • Solving word problems
UNIT 7	**DATA ANALYSIS** • Describing Data • Organizing Data • Graphing Data and Making Predictions	**PLANE GEOMETRY** • Perimeter and Area • Symmetry and Reflections • Other Transformations	**RADICAL EXPRESSIONS** • Rational and irrational numbers • Operations with radicals • Irrational roots • Radical equations
UNIT 8	**GEOMETRY** • Basic Geometry • Classifying Polygons • Transformations	**MEASURE OF SOLID FIGURES** • Surface Area • Solid Figures • Volume • Volume of Composite Figures	**GRAPHING** • Equations of two variables • Graphing lines • Graphing inequalities • Equations of lines
UNIT 9	**MEASUREMENT AND AREA** • Perimeter • Area • The Pythagorean Theorem	**DATA ANALYSIS** • Collecting and Representing Data • Central Tendency and Dispersion • Frequency and Histograms • Box-and-Whisker Plots • Scatter Plots	**SYSTEMS** • Graphical solution • Algebraic solutions • Determinants • Word problems
UNIT 10	**SURFACE AREA AND VOLUME** • Solids • Prisms • Cylinders	**PROBABILITY** • Outcomes • Permutations and Combinations • Probability and Odds • Independent and Dependent Events	**QUADRATIC EQUATIONS AND REVIEW** • Solving quadratic equations • Equations and inequalities • Polynomials and factors • Radicals and graphing

MATH SCOPE & SEQUENCE

Grade 10	Grade 11	Grade 12	
A MATHEMATICAL SYSTEM • Points, lines, and planes • Definition of definitions • Geometric terms • Postulates and theorems	**SETS, STRUCTURE, AND FUNCTION** • Properties and operations of sets • Axioms and applications • Relations and functions • Algebraic expressions	**RELATIONS AND FUNCTIONS** • Relations and functions • Rules of correspondence • Notation of functions • Types of functions	UNIT 1
PROOFS • Logic • Reasoning • Two-column proof • Paragraph proof	**NUMBERS, SENTENCES, & PROBLEMS** • Order and absolute value • Sums and products • Algebraic sentences • Number and motion problems	**SPECIAL FUNCTIONS** • Linear functions • Second-degree functions • Polynomial functions • Other functions	UNIT 2
ANGLES AND PARALLELS • Definitions and measurement • Relationships and theorems • Properties of parallels • Parallels and polygons	**LINEAR EQUATIONS & INEQUALITIES** • Graphs • Equations • Systems of equations • Inequalities	**TRIGONOMETRIC FUNCTIONS** • Definition • Equation of functions • Trigonometric tables • Special angles	UNIT 3
CONGRUENCY • Congruent triangles • Corresponding parts • Inequalities • Quadrilaterals	**POLYNOMIALS** • Multiplying polynomials • Factoring • Operations with polynomials • Variations	**CIRCULAR FUNCTIONS & GRAPHS** • Circular functions & special angles • Graphs of sine and cosine • Amplitude and period • Phase shifts	UNIT 4
SIMILAR POLYGONS • Ratios and proportions • Definition of similarity • Similar polygons and triangles • Right triangle geometry	**RADICAL EXPRESSIONS** • Multiplying and dividing fractions • Adding and subtracting fractions • Equations with fractions • Applications of fractions	**IDENTITIES AND FUNCTIONS** • Reciprocal relations • Pythagorean relations • Trigonometric identities • Sum and difference formulas	UNIT 5
CIRCLES • Circles and spheres • Tangents, arcs, and chords • Special angles in circles • Special segments in circles	**REAL NUMBERS** • Rational and irrational numbers • Laws of Radicals • Quadratic equations • Quadratic formula	**TRIGONOMETRIC FUNCTIONS** • Trigonometric functions • Law of cosines • Law of sines • Applied problems	UNIT 6
CONSTRUCTION AND LOCUS • Basic constructions • Triangles and circles • Polygons • Locus meaning and use	**QUADRATIC RELATIONS & SYSTEMS** • Distance formulas • Conic sections • Systems of equations • Application of conic sections	**INVERSE TRIGONOMETRIC FUNCTIONS** • Inverse functions • Graphing polar coordinates • Converting polar coordinates • Graphing polar equations	UNIT 7
AREA AND VOLUME • Area of polygons • Area of circles • Surface area of solids • Volume of solids	**EXPONENTIAL FUNCTIONS** • Exponents • Exponential equations • Logarithmic functions • Matrices	**QUADRATIC EQUATIONS** • Conic sections • Circle and ellipse • Parabola and hyperbola • Transformations	UNIT 8
COORDINATE GEOMETRY • Ordered pairs • Distance • Lines • Coordinate proofs	**COUNTING PRINCIPLES** • Progressions • Permutations • Combinations • Probability	**PROBABILITY** • Random experiments & probability • Permutations • Combinations • Applied problems	UNIT 9
REVIEW • Proof and angles • Polygons and circles • Construction and measurement • Coordinate geometry	**REVIEW** • Integers and open sentences • Graphs and polynomials • Fractions and quadratics • Exponential functions	**CALCULUS** • Mathematical induction • Functions and limits • Slopes of functions • Review	UNIT 10

STRUCTURE OF THE LIFEPAC CURRICULUM

The LIFEPAC curriculum is conveniently structured to provide one Teacher's Guide containing teacher support material with answer keys and ten student worktexts for each subject at grade levels two through twelve. The worktext format of the LIFEPACs allows the student to read the textual information and complete workbook activities all in the same booklet. The easy-to-follow LIFEPAC numbering system lists the grade as the first number(s) and the last two digits as the number of the series. For example, the Language Arts LIFEPAC at the 6th grade level, 5th book in the series would be LAN0605.

Each LIFEPAC is divided into 3 to 5 sections and begins with an introduction or overview of the booklet as well as a series of specific learning objectives to give a purpose to the study of the LIFEPAC. The introduction and objectives are followed by a vocabulary section which may be found at the beginning of each section at the lower levels or in the glossary at the high school level. Vocabulary words are used to develop word recognition and should not be confused with the spelling words introduced later in the LIFEPAC. The student should learn all vocabulary words before working the LIFEPAC sections to improve comprehension, retention, and reading skills.

Each activity or written assignment in grades 2 through 12 has a number for easy identification, such as 1.1. The first number corresponds to the LIFEPAC section and the number to the right of the decimal is the number of the activity.

Teacher checkpoints, which are essential to maintain quality learning, are found at various locations throughout the LIFEPAC. The teacher should check 1) neatness of work and penmanship, 2) quality of understanding (tested with a short oral quiz), 3) thoroughness of answers (complete sentences and paragraphs, correct spelling, etc.), 4) completion of activities (no blank spaces), and 5) accuracy of answers as compared to the answer key (all answers correct).

The self test questions in grades 2 through 12 are also number coded for easy reference. For example, 2.015 means that this is the 15th question in the self test of Section 2. The first number corresponds to the LIFEPAC section, the zero indicates that it is a self test question, and the number to the right of the zero the question number.

The LIFEPAC test is packaged at the centerfold of each LIFEPAC. It should be removed and put aside before giving the booklet to the student for study.

Answer and test keys in grades 2 through 12 have the same numbering system as the LIFEPACs. The student may be given access to the answer keys (not the test keys) under teacher supervision so that he can score his own work.

A thorough study of the LIFEPAC Overview by the teacher before instruction begins is essential to the success of the student. The teacher should become familiar with expected skill mastery and understand how these grade-level skills fit into the overall skill development of the curriculum. The teacher should also preview the objectives that appear at the beginning of each LIFEPAC for additional preparation and planning.

TEST SCORING AND GRADING

Answer keys and test keys give examples of correct answers. They convey the idea, but the student may use many ways to express a correct answer. The teacher should check for the essence of the answer, not for the exact wording. Many questions are high level and require thinking and creativity on the part of the student. Each answer should be scored based on whether or not the main idea written by the student matches the model example. "Any Order" or "Either Order" in a key indicates that no particular order is necessary to be correct.

Most self tests and LIFEPAC tests at the lower elementary levels are scored at 1 point per answer; however, the upper levels may have a point system awarding 2 to 5 points for various answers or questions. Further, the total test points will vary; they may not always equal 100 points. They may be 78, 85, 100, 105, etc.

Example 1

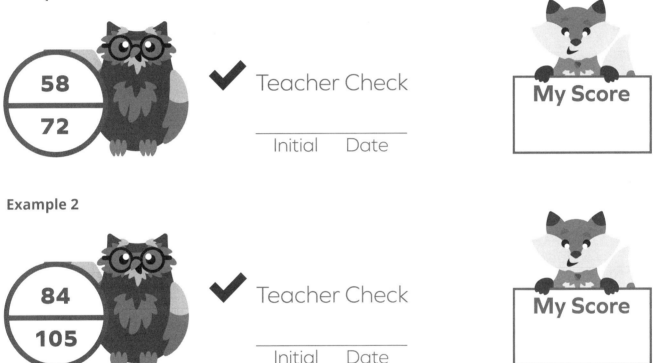

Example 2

A score box similar to ex. 1 above is located at the end of each self test and on the front of the LIFEPAC test. The bottom score, 72, represents the total number of points possible on the test. The upper score, 58, represents the number of points your student will need to receive an 80% or passing grade. If you wish to establish the exact percentage that your student has achieved, find the total points of his correct answers and divide it by the bottom number (in this case 72). For example, if your student has a point total of 65, divide 65 by 72 for a grade of 90%. Referring to ex. 2, on a test with a total of 105 possible points, the student would have to receive a minimum of 84 correct points for an 80% or passing grade. If your student has received 93 points, simply divide the 93 by 105 for a percentage grade of 89%. Students who receive a score below 80% should review the LIFEPAC and retest using the appropriate Alternate Test found in the Teacher's Guide.

The following is a guideline to assign letter grades for completed LIFEPACs based on a maximum total score of 100 points.

Example:

LIFEPAC Test	=	60% of the Total Score (or percent grade)
Self Test	=	25% of the Total Score (average percent of self tests)
Reports	=	10% or 10* points per LIFEPAC
Oral Work	=	5% or 5* points per LIFEPAC

*Determined by the teacher's subjective evaluation of the student's daily work.

Example:

LIFEPAC Test Score	=	92%	92 × .60	=	55 points
Self Test Average	=	90%	90 × .25	=	23 points
Reports				=	8 points
Oral Work				=	4 points
TOTAL POINTS				=	90 points

Grade Scale based on point system:

100 – 94	=	A
93 – 86	=	B
85 – 77	=	C
76 – 70	=	D
Below 70	=	F

TEACHER HINTS AND STUDYING TECHNIQUES

LIFEPAC activities are written to check the level of understanding of the preceding text. The student may look back to the text as necessary to complete these activities; however, a student should never attempt to do the activities without reading (studying) the text first. Self tests and LIFEPAC tests are never open book tests.

Language arts activities (skill integration) often appear within other subject curriculum. The purpose is to give the student an opportunity to test his skill mastery outside of the context in which it was presented.

Writing complete answers (paragraphs) to some questions is an integral part of the LIFEPAC curriculum in all subjects. This builds communication and organization skills, increases understanding and retention of ideas, and helps enforce good penmanship. Complete sentences should be encouraged for this type of activity. Obviously, single words or phrases do not meet the intent of the activity, since multiple lines are given for the response.

Review is essential to student success. Time invested in review where review is suggested will be time saved in correcting errors later. Self tests, unlike the section activities, are closed book. This procedure helps to identify weaknesses before they become too great to overcome. Certain objectives from self tests are cumulative and test previous sections; therefore, good preparation for a self test must include all material studied up to that testing point.

The following procedure checklist has been found to be successful in developing good study habits in the LIFEPAC curriculum.

1. Read the introduction and Table of Contents.
2. Read the objectives.
3. Recite and study the entire vocabulary (glossary) list.
4. Study each section as follows:
   a. Read the introduction and study the section objectives.
   b. Read all the text for the entire section, but answer none of the activities.
   c. Return to the beginning of the section and memorize each vocabulary word and definition.
   d. Reread the section, complete the activities, check the answers with the answer key, correct all errors, and have the teacher check.
   e. Read the self test but do not answer the questions.
   f. Go to the beginning of the first section and reread the text and answers to the activities up to the self test you have not yet done.
   g. Answer the questions to the self test without looking back.
   h. Have the self test checked by the teacher.
   i. Correct the self test and have the teacher check the corrections.
   j. Repeat steps a–i for each section.
5. Use the **SQ3R** method to prepare for the LIFEPAC test.
   > **S**can the whole LIFEPAC.
   > **Q**uestion yourself on the objectives.
   > **R**ead the whole LIFEPAC again.
   > **R**ecite through an oral examination.
   > **R**eview weak areas.
6. Take the LIFEPAC test as a closed book test.
7. LIFEPAC tests are administered and scored under direct teacher supervision. Students who receive scores below 80% should review the LIFEPAC using the **SQ3R** study method and take the Alternate Test located in the Teacher's Guide. The final test grade may be the grade on the Alternate Test or an average of the grades from the original LIFEPAC test and the Alternate Test.

GOAL SETTING AND SCHEDULES

Each school must develop its own schedule, because no single set of procedures will fit every situation. The following is an example of a daily schedule that includes the five LIFE-PAC subjects as well as time slotted for special activities.

Possible Daily Schedule

8:15 – 8:25	Pledges, prayer, songs, devotions, etc.
8:25 – 9:10	Bible
9:10 – 9:55	Language Arts
9:55 – 10:15	Recess (juice break)
10:15 – 11:00	Math
11:00 – 11:45	History & Geography
11:45 – 12:30	Lunch, recess, quiet time
12:30 – 1:15	Science
1:15 –	Drill, remedial work, enrichment*

*Enrichment: *Computer time, physical education, field trips, fun reading, games and puzzles, family business, hobbies, resource persons, guests, crafts, creative work, electives, music appreciation, projects.*

Basically, two factors need to be considered when assigning work to a student in the LIFE-PAC curriculum.

The first is time. An average of 45 minutes should be devoted to each subject, each day. Remember, this is only an average. Because of extenuating circumstances a student may spend only 15 minutes on a subject one day and the next day spend 90 minutes on the same subject.

The second factor is the number of pages to be worked in each subject. A single LIFEPAC is designed to take 3 to 4 weeks to complete. Allowing about 3 to 4 days for LIFEPAC introduction, review, and tests, the student has approximately 15 days to complete the LIFEPAC pages. Simply take the number of pages in the LIFEPAC, divide it by 15 and you will have the number of pages that must be completed on a daily basis to keep the student on schedule. For example, a LIFEPAC containing 45 pages will require 3 completed pages per day. Again, this is only an average. While working a 45-page LIFEPAC, the student may complete only 1 page the first day if the text has a lot of activities or reports, but go on to complete 5 pages the next day.

Long-range planning requires some organization. Because the traditional school year originates in the early fall of one year and continues to late spring of the following year, a calendar should be devised that covers this period of time. Approximate beginning and completion dates can be noted on the calendar as well as special occasions such as holidays, vacations and birthdays. Since each LIFEPAC takes 3 to 4 weeks or eighteen days to complete, it should take about 180 school days to finish a set of ten LIFEPACs. Starting at the beginning school date, mark off eighteen school days on the calendar and that will become the targeted completion date for the first LIFEPAC. Continue marking the calendar until you have established dates for the remaining nine LIFEPACs making adjustments for previously noted holidays and vacations. If all five subjects are being used, the ten established target dates should be the same for the LIFEPACs in each subject.

TEACHING SUPPLEMENTS

The sample weekly lesson plan and student grading sheet forms are included in this section as teacher support materials and may be duplicated at the convenience of the teacher.

The student grading sheet is provided for those who desire to follow the suggested guidelines for assignment of letter grades as previously discussed. The student's self test scores should be posted as percentage grades. When the LIFEPAC is completed, the teacher should average the self test grades, multiply the average by .25, and post the points in the box marked self test points. The LIFEPAC percentage grade should be multiplied by .60 and posted. Next, the teacher should award and post points for written reports and oral work. A report may be any type of written work assigned to the student whether it is a LIFEPAC or additional learning activity. Oral work includes the student's ability to respond orally to questions which may or may not be related to LIFEPAC activities or any type of oral report assigned by the teacher. The points may then be totaled and a final grade entered along with the date that the LIFEPAC was completed.

The Student Record Book which was specifically designed for use with the Alpha Omega curriculum provides space to record weekly progress for one student over a nine-week period as well as a place to post self test and LIFEPAC scores. The Student Record Books are available through the current Alpha Omega catalog; however, unlike the enclosed forms these books are not for duplication and should be purchased in sets of four to cover a full academic year.

WEEKLY LESSON PLANNER

Week of: _____

	Subject	Subject	Subject	Subject
Monday				
Tuesday				
Wednesday				
Thursday				
Friday				

WEEKLY LESSON PLANNER

Week of:

	Subject	Subject	Subject	Subject
Monday				
Tuesday	Subject	Subject	Subject	Subject
Wednesday	Subject	Subject	Subject	Subject
Thursday	Subject	Subject	Subject	Subject
Friday	Subject	Subject	Subject	Subject

Student Name _____   Year _____

Bible

LP	Self Test Scores by Sections 1	2	3	4	5	Self Test Points	LIFEPAC Test	Oral Points	Report Points	Final Grade	Date
01											
02											
03											
04											
05											
06											
07											
08											
09											
10											

History & Geography

LP	Self Test Scores by Sections 1	2	3	4	5	Self Test Points	LIFEPAC Test	Oral Points	Report Points	Final Grade	Date
01											
02											
03											
04											
05											
06											
07											
08											
09											
10											

Language Arts

LP	Self Test Scores by Sections 1	2	3	4	5	Self Test Points	LIFEPAC Test	Oral Points	Report Points	Final Grade	Date
01											
02											
03											
04											
05											
06											
07											
08											
09											
10											

Student Name _____   Year _____

Math

LP	Self Test Scores by Sections					Self Test Points	LIFEPAC Test	Oral Points	Report Points	Final Grade	Date
	1	2	3	4	5						
01											
02											
03											
04											
05											
06											
07											
08											
09											
10											

Science

LP	Self Test Scores by Sections					Self Test Points	LIFEPAC Test	Oral Points	Report Points	Final Grade	Date
	1	2	3	4	5						
01											
02											
03											
04											
05											
06											
07											
08											
09											
10											

Spelling/Electives

LP	Self Test Scores by Sections					Self Test Points	LIFEPAC Test	Oral Points	Report Points	Final Grade	Date
	1	2	3	4	5						
01											
02											
03											
04											
05											
06											
07											
08											
09											
10											

INSTRUCTIONS FOR MATH

The first grade Teacher's Guides of the LIFEPAC curriculum are designed to provide a step-by-step procedure that will help the teacher prepare for and present each lesson effectively. In the early LIFEPACs, the teacher should read the directions and any other sentences to the children. However, as the school year progresses, the student should be encouraged to begin reading and following his own instructional material in preparation for the independent study approach that begins at the second grade level.

The remainder of the Teacher's Guide includes the following teacher aids:

1) Introduction of Skills

For each unit:
2) Materials Needed
3) Objectives
4) Teacher Instruction
5) Answer Keys
6) Alternate Tests

After the last unit:
7) Math Terms Glossary
8) Conversion Charts

The Introduction of Skills is a more detailed overview of skills than that presented in the *Scope and Sequence*. The Math Terms includes a glossary of math terms and a table of measurements. The Teacher Instruction Pages contain guidelines for teaching each lesson. Additional learning activities provide opportunities for problem solving, encourage the student's interest in learning, and may be used as a reward for good study habits.

Math is a subject that requires skill mastery. But skill mastery needs to be applied toward active student involvement. The Teacher Instruction Pages list the required or suggested materials used in the LIFEPAC lessons. These materials include items generally available in the school or home. Pencils, paper, crayons, scissors, paste and/or glue stick are materials used on a regular basis. Construction paper, beads, buttons, and beans can be used for counting, sets, grouping, fractions, and patterning. Measurements require measuring cups, rulers, and empty containers. Boxes and similar items help in the study of solid shapes.

Any workbook assignment that can be supported by a real-world experience will enhance the student's ability for problem solving. There is an infinite challenge for the teacher to provide a meaningful environment for the study of math. It is a subject that requires constant assessment of student progress. Do not leave the study of math in the classroom.

MATH 100 INTRODUCTION OF SKILLS

Introduction of Skills is a quick reference guide for the teacher who may be looking for a rule or explanation that applies to a particular skill or to find where or when certain skills are introduced in the LIFEPACs. The first number after the skill identifies the LIFEPAC, and the second number identifies the section.

CONCEPT	LIFEPAC	SECTION
Addition		
facts to 9	101	3
facts to 10	102	1
1-digit number added to 10	102	3
facts to 18	104	1
3 numbers, 1 digit	105	1
2 numbers, 2 digits	106	3
checking answers	105	1
3 numbers, 2 digits	109	1
Calendar	103	4
Count		
to 99	101	1
to 100	103	4
to 200	107	1
Directions		
north, east, south, west	110	3
Estimation		
size and weight	102	4
numbers	107	5
Even and odd numbers	104	2
Families of facts		
addition and subtraction	109	1
Fractions		
$\frac{1}{2}$ of an object, of a set	103	3
$\frac{1}{4}$ of an object, of a set	105	3
Graphs (Charts)		
posting data	105	4
	107	5
	109	3
Measurements		
objects big and little	101	4
objects greater than, less than	101	4
long and short	101	4
dozen	105	3
ruler—inch	101	4
ruler—one-half inch	108	2
weight	102	4
Money		
pennies, dimes	104	4
nickels	106	4

CONCEPT	LIFEPAC	SECTION
Number line	101	1
Number order		
before and after to 99	101	2
bigger and smaller to 99	101	2
before and after to 100	103	5
before and after to 200	107	3
greater than, less than to 100	105	2
greater than, less than to 200	107	1
closest multiple of 10	109	3
	110	1
Number sentences	103	2
Number words		
zero to ten	102	5
to twenty	104	3
to ninety-nine	106	2
Operation symbols		
+, −, =	102	1
≠	103	2
>, <	106	2
Ordinal numbers		
to tenth	102	5
Place value		
for ones	102	3
for tens	102	3
for hundreds	107	2
Problem solving		
estimation	102	4
how many facts equal a number	102	5
sensible answers	109	4
Sequencing and number patterns	102	5
Shapes		
flat	101	4
solid	102	4
Skip counting		
by 10's	101	1
	104	2
by 2's	104	2
by 5's	106	1
objects by grouping	108	5

MATH 100 INTRODUCTION OF SKILLS

CONCEPT	LIFEPAC	SECTION
Story problems		
oral problems	101	4
	103	3
oral/written	103	5
written	104	4
Subtraction		
facts to 9	101	3
facts to 10	102	2
facts to 12	107	3
facts to 18	108	1,3
1-digit from 10's n/b*	109	4
2 numbers, 2 digits n/b*	109	4
Time		
to hour	103	4
to half-hour	105	3
to quarter-hour	106	4
to 5 minutes	108	2
A.M., P.M.	107	4
Write numbers		
to 99	101	1–3
to 200	107	1
Zero as a place holder	107	2

CONCEPT	LIFEPAC	SECTION

*n/b = no borrowing

MATH 101

Unit 1: Number Order, Add/Subtract

NUMBER ORDER, ADD/SUBTRACT
MATH 101

Alpha Omega
PUBLICATIONS

Author:
Carol Bauler, B.A.

Editor:
Alan Christopherson, M.S.

Media Credits:
Page 1: © wenchiawang, iStock, Thinkstock

804 N. 2nd Ave. E.
Rock Rapids, IA 51246-1759

|i

PAGE 1: NUMBER ORDER, ADD/SUBTRACT

MATERIALS NEEDED

• pencils

Concepts:

purpose of LIFEPAC, objectives

Teacher Goals:

To teach the children to know what is expected of the student in the LIFEPAC and to write first and last names correctly.

Teaching Page 1:

Turn to page 1. Point to the title and read it aloud. Allow time for the children to look through the LIFEPAC. Write the word *Objectives* on the board and have the children find the word on the page. Explain that the objectives tell the things the students will be expected to do in the LIFEPAC. Read each one and have the children repeat them as they run their fingers under the sentence from left to right. Talk about the objectives so that the children will understand what they will be doing. Have each child write his name on the line.

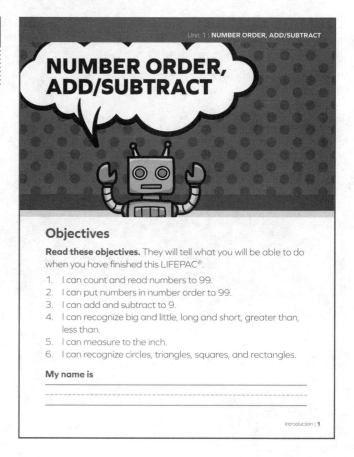

Unit 1 | **NUMBER ORDER, ADD/SUBTRACT**

NUMBER ORDER, ADD/SUBTRACT

Objectives

Read these objectives. They will tell what you will be able to do when you have finished this LIFEPAC®.

1. I can count and read numbers to 99.
2. I can put numbers in number order to 99.
3. I can add and subtract to 9.
4. I can recognize big and little, long and short, greater than, less than.
5. I can measure to the inch.
6. I can recognize circles, triangles, squares, and rectangles.

My name is

Introduction | 1

1.   COUNT TO 99

PAGE 2: NUMBERS TO 9

MATERIALS NEEDED

- pencils
- paper
- twelve strips of colored paper (all one color), one inch by five inches in size, pasted or glued to cardboard (cereal boxes work well), or any objects that may be used for counting (Popsicle sticks, beads, buttons, blocks)—one set for each student

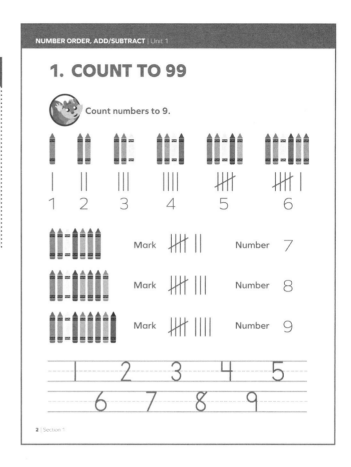

NUMBER ORDER, ADD/SUBTRACT | Unit 1

1. COUNT TO 99

Count numbers to 9.

2 | Section 1

Concept:

counting numbers to 9

Teacher Goals:

To teach the children to make tally marks for numbers to 9 and to count numbers to 9.

Teaching Page 2:

Place the objects for counting in front of the students. Ask the children to select the amount of strip(s) from the pile that represents the number *1* and to make a tally mark for *1* on a piece of paper. Continue doing this through the number *9*. Be sure students form the tally marks for number *5* correctly.

Turn to page 2. Ask the children to point to the first crayon and to the tally mark below it. Ask them to read the number *1*. Continue in this manner through *9*. Have the children read the numbers *1* through *9* at the bottom of the page and trace the numbers as they read them. Tell the students to close their books and ask them to use the objects for counting to count to *9* independently.

PAGE 3: WRITE NUMBERS TO 9

MATERIALS NEEDED

- pencils
- objects for counting

Concept:

writing numbers from 0 to 9

Teacher Goals:

To teach the children to recognize and write the number symbol for zero and to write numbers 0 to 9.

Teaching Page 3:

Place *nine* objects in front of the students and ask them to count them. Remove the objects from the table. Ask: "Now how many objects?" Elicit responses of *none*, *nothing*, and *zero*. Ask the children if they know how to write a symbol for *none*, *nothing*, or *zero*. Write *0* on the board.

Turn to page 3. Tell the children to look at the first column and read the numbers *1* through *9* aloud. Ask them if they know the name of the last number symbol on the page. Explain to the children that this number symbol stands for *none*, *nothing* or *zero*. Have the students trace the numbers on the page with their fingers. Then have them write each number five times. Place the objects for counting in front of the students. Select numbers from *0* through *9* in random order and have the students show how many objects represent the number. Continue this exercise until students are competent in their selections.

Unit 1 | **NUMBER ORDER, ADD/SUBTRACT**

Write the numbers.

1
2
3
4
5
6
7
8
9
0

Section 1 | **3**

PAGES 4 & 5: COUNT NUMBERS TO 19

MATERIALS NEEDED

- pencils
- objects for counting

Concept:

counting numbers to 19

Teacher Goal:

To teach the children to tell "how many" by counting to nineteen.

Teaching Pages 4 and 5:

Draw a number line for *0* through *19* on the board. Introduce the numbers *10*, *11*, *12*, *13*, *14*, *15*, *16*, *17*, *18*, and *19* as the numbers that follow *9*. Have the children say the numbers aloud several times. Place groups of objects for counting from *0* through *19* in front of the students in random order. Have them count the objects and then write the number on the paper. Do this several times until the children understand what you are asking them to do and are familiar with the numbers through *19*.

Turn to page 4. Point to the word *count* and have the children read it aloud. Tell the children to point to the leaves and count them aloud. Have them trace the numbers *10* through *19* at the bottom of the page. Tell the children to close their books and use the objects for counting to count to *19*.

Turn to page 5. Have the children write each number five times.

PAGE 6: COUNT TO 99

MATERIALS NEEDED

- pencils
- paper
- objects for counting (40 objects)

Concepts:

sets of 10, counting to 99

Teacher Goals:

To teach the children to group objects in sets of 10 and to count to 99.

Teaching Page 6:

Introduce the word *set* to the students. Explain to them that a set of objects is the same as a group of objects. Tell the students to use the objects for counting to make a set of *ten* objects. Have them count the set aloud and write *10* on a piece of paper. Have them make another set of *ten* objects. Keep the two sets separate but have the students use them to count to *19*. Introduce the number *20* to them. Write *20* on the board and then ask the students to write the number. Tell them to use the objects for counting to make another set of *ten* objects. Introduce the number *30*. Ask the children to write this number on paper. Have them use the three sets to count to *30*.

Turn to page 6. Read the instructions on the page to the children. Tell them that they can count to *99*. Start at the first row of numbers and have the students count aloud to *10*. Tell the children that this represents their first set of ten objects. Continue on the second row of numbers and have the students count aloud. Call attention to the number *20*. Tell them that this represents their second set of *ten* objects. Go on with the third row, calling attention to the number *30*.

Ask the students to continue counting from *31* through *40*. Have them pull an object for counting for each number that they count. Ask them if they now have another set of *ten*. (yes) Continue counting aloud with the students until they reach *99*. Call attention to the *50, 60, 70, 80,* and *90*. Have the children count again from *0* through *99*, pointing at each number as they say it aloud.

Count to 99.

0	1	2	3	4	5	6	7	8	9
10	11	12	13	14	15	16	17	18	19
20	21	22	23	24	25	26	27	28	29
30	31	32	33	34	35	36	37	38	39
40	41	42	43	44	45	46	47	48	49
50	51	52	53	54	55	56	57	58	59
60	61	62	63	64	65	66	67	68	69
70	71	72	73	74	75	76	77	78	79
80	81	82	83	84	85	86	87	88	89
90	91	92	93	94	95	96	97	98	99

6 | Section 1

PAGE 7: COUNT TO 99

MATERIALS NEEDED

- pencils
- chart of numbers to 99

Concept:

counting to 99

Teacher Goal:

To teach the children to count to 99 by writing the missing numbers.

Teaching Page 7:

A chart of numbers similar to the chart on page 6 should be prepared for student use. It may be taken directly from the LIFEPAC and pasted onto cardboard or copied to cardboard.

Turn to page 7. Read the instructions at the top of the page. Ask the children what this chart looks like and what is wrong with it. (It is like the chart they studied on page 6 but some numbers are missing.) Tell the students that they should write the numbers to *99* by filling in the empty boxes. Ask the students to say the first missing number aloud and then write it in the box. (0) Students who are able may complete the page independently. Other children may need to refer to the chart of numbers. When the page is completed have the children read the chart aloud.

The students should prepare for the Self Test. Ask the students to look over and read the Self Test but they should not write the answers to any questions. After looking over the Self Test the students should go to the beginning of the unit and reread the text and review the answers to the activities up to the Self Test.

The students are to complete the Self Test the next school day. This should be done under regular test conditions without allowing the students to look back. A good idea is to clip the pages together before the test.

Unit 1 | **NUMBER ORDER, ADD/SUBTRACT**

Write Numbers to 99

Count to 99. Write the number in the ☐ .

0	1	2	3	4	5	6	7	8	9
10	11	12	13	14	15	16	17	18	19
20	21	22	23	24	25	26	27	28	29
30	31	32	33	34	35	36	37	38	39
40	41	42	43	44	45	46	47	48	49
50	51	52	53	54	55	56	57	58	59
60	61	62	63	64	65	66	67	68	69
70	71	72	73	74	75	76	77	78	79
80	81	82	83	84	85	86	87	88	89
90	91	92	93	94	95	96	97	98	99

Before you take the Self Test, study what you have read and done. The Self Test will check what you remember.

Section 1 | **7**

PAGE 8: SELF TEST 1

MATERIALS NEEDED

- pencils
- chart of numbers to 99

Concept:

counting to 99

Teacher Goal:

To teach the children to learn to check their progress periodically.

Teaching Page 8:

Turn to page 8. Read the directions to the children. Have the children repeat them after you while running their fingers under the sentence being read. Be sure the children understand what they are to do. If necessary, students may use the chart of numbers. Let the children complete the page. You may repeat the directions but give no other help.

Do not have the children check their own work. Check it as soon as you can, and go over it with each child. Show him where he did well and where he needs extra help. Students having difficulty counting to *99* should practice counting using the chart of numbers on a daily basis until the skill is mastered.

2. NUMBER ORDER TO 99

PAGE 9: COUNT TO 99

MATERIALS NEEDED

- pencils
- paper
- chart of numbers to 99
- crayons—colors red, yellow, green, blue, brown, purple, orange, black, white, and pink

Concept:

counting to 99

Teacher Goal:

To teach the children to count to 99 by going from dot to dot.

Teaching Page 9:

On a piece of paper, make a series of twelve dots in the shape of a box. Number the dots from *1* to *12*. Tell the students to connect the dots with straight lines beginning with *1* and counting to *12*. Ask the students to name the shape that they have made by connecting the dots. (square, box)

2. NUMBER ORDER TO 99

Draw a line from dot to dot. Start at 1. Stop at 99.

The camel has a long, long neck,
And pads upon his feet.
He carries food inside his hump,
It makes a cozy seat.

Color.

Section 2 | **9**

Turn to page 9 and read the instructions aloud. Help the children begin by pointing out the *1* and the *2*. Ask them to find the next number. Allow the students to complete the exercise as independently as possible. They may use the chart of numbers if necessary. When the dot-to-dot is completed, read the poem aloud with the children. Talk with them about where a camel lives and how it helps us.

Place the crayons in front of the students and have them identify each crayon as you say the name of the color. Note should be taken of any one of the colors that the students cannot identify and time should be spent until all of the colors listed above are mastered by each student. This can be done during free time or when working at coloring in other subjects. Have the students complete page 9 using their crayons for coloring.

PAGE 10: BIGGER THAN, SMALLER THAN

MATERIALS NEEDED

- pencils
- objects for counting
- chart of numbers to 99

Concept:

numbers bigger than and smaller than to 99

Teacher Goal:

To teach the children to identify numbers by finding bigger than and smaller than to 99.

Teaching Page 10:

Place sets of *2* objects and *8* objects in front of the students. Ask them to identify which set is *bigger* and which set is *smaller*. Do the same for sets of *8* and *15*, and sets of *15* and *22*. Use the chart of numbers and identify the sets (numbers) on the chart. Have the students identify the pattern of *bigger* and *smaller* numbers. Find the number *25* on the chart. Point to several other numbers and ask if they are *bigger* or *smaller* than *25*. Show where they are on the chart in relation to the *25*.

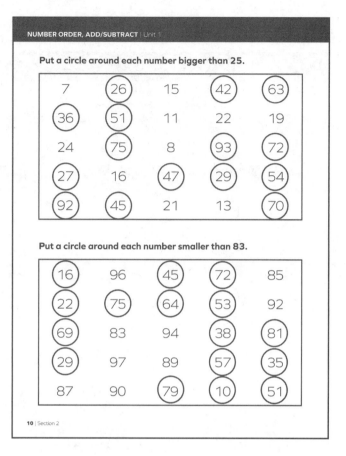

Turn to page 10 and read the first set of directions aloud. Emphasize the word *bigger*. Let the children use the chart of numbers to complete the exercise. Read the second set of directions to the students. Emphasize the word *smaller*. The children may use the chart of numbers to complete the exercise. This type of exercise may be continued using the chart of numbers. Select a number at random and ask the students to say numbers that are *bigger than* or *smaller than*. Continue as needed.

PAGE 11: MISSING NUMBERS

MATERIALS NEEDED

- pencils
- chart of numbers to 99

Concept:

writing missing numbers to 99

Teacher Goal:

To teach the children to write the missing numbers by counting to 99.

Teaching Page 11:

Turn to page 11. Read the directions aloud to the students. Have them point to the first row of numbers and ask them to say the first three numbers aloud. (six-seven-eight) Have them find *6*, *7*, and *8* on the chart of numbers. Ask them to identify the next number on the chart (*9*) and write it on the blank line. Ask them to find the next number and write it on the line. Continue until the first line is completed. Then have the students read the line aloud. Have the students point to the second row of numbers. Be sure they understand that this is not a continuation of the first row. Follow the same procedure as the first row. Continue in the same manner, but monitor each student's work carefully row by row to be sure he understands the concept.

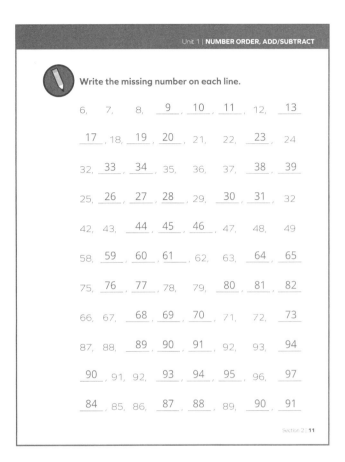

Unit 1 | **NUMBER ORDER, ADD/SUBTRACT**

Write the missing number on each line.

6, 7, 8, __9__, __10__, __11__, 12, __13__

__17__, 18, __19__ __20__, 21, 22, __23__, 24

32, __33__ __34__, 35, 36, 37, __38__ __39__

25, __26__ __27__ __28__, 29, __30__ __31__, 32

42, 43, __44__ __45__ __46__, 47, 48, 49

58, __59__ __60__ __61__, 62, 63, __64__ __65__

75, __76__ __77__, 78, 79, __80__ __81__ __82__

66, 67, __68__ __69__ __70__, 71, 72, __73__

87, 88, __89__ __90__ __91__, 92, 93, __94__

__90__, 91, 92, __93__ __94__ __95__, 96, __97__

__84__, 85, 86, __87__ __88__, 89, __90__ __91__

Section 2 | **11**

PAGE 12: BEFORE AND AFTER

MATERIALS NEEDED

- pencils
- objects for counting
- twenty pieces of cardboard, two inches by three inches—write the numbers 0 through 9 on each of two pieces
- chart of numbers to 99

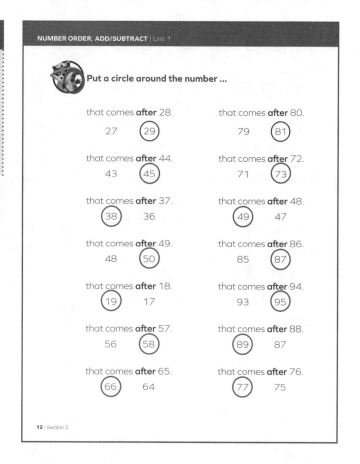

Concept:

writing numbers after to 99

Teacher Goal:

To teach the children to select the numbers after to 99.

Teaching Page 12:

Place one set of number symbol cards in front of the students and have them arrange the cards in order from *0* to *9*. Select the number card *8* and ask the children to find the number card that comes *before* and *after*. (*7, 9*) Continue this exercise with other numbers until the students understand that you mean the number immediately *before* and *after*.

Turn to the number chart. Selecting any number up to *98* at random, point to the number on the chart. Have the student read the number that comes *before* and the number that comes *after*. Ask the student which number is bigger. Always use objects for counting to illustrate if a student gives an incorrect answer.

Turn to page 12. Read the instructions at the top of the page and point to the word *after*. Have the students read the word aloud. Point to the *28* and have the students find it on their number chart. Ask them if *29* is the correct answer. Point to the *80* and have the students find it on their number chart. Ask them if *81* is the correct answer. Emphasize that they should circle the number *after*. Allow them to complete the page independently.

PAGE 13: BEFORE AND AFTER

MATERIALS NEEDED

- pencil
- number symbol cards from page 12
- chart of numbers to 99

Concept:

writing numbers before to 99

Teacher Goal:

To teach the children to select the numbers before to 99.

Teaching Page 13:

Repeat the exercise from page 12 using the number symbol cards to identify numbers *before* and *after*.

Turn to page 13 and read the instructions. Work the two examples with the students and then allow them to complete the page independently. Be sure they understand that, on this page, they are to circle the number *before*.

The students should prepare for the Self Test. Ask the students to look over and read the Self Test but they should not write the answers to any questions. After looking over the Self Test the students should go to the beginning of the unit and reread the text and review the answers to the activities up to the Self Test.

The students are to complete the Self Test the next school day. This should be done under regular test conditions without allowing the students to look back. A good idea is to clip the pages together before the test.

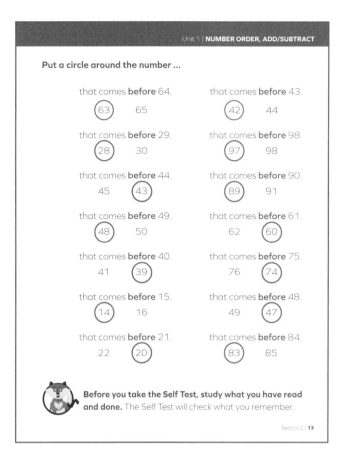

Unit 1 | **NUMBER ORDER, ADD/SUBTRACT**

Put a circle around the number ...

that comes **before** 64.
(63)   65

that comes **before** 43.
(42)   44

that comes **before** 29.
(28)   30

that comes **before** 98.
(97)   98

that comes **before** 44.
45   (43)

that comes **before** 90.
(89)   91

that comes **before** 49.
(48)   50

that comes **before** 61.
62   (60)

that comes **before** 40.
41   (39)

that comes **before** 75.
76   (74)

that comes **before** 15.
(14)   16

that comes **before** 48.
49   (47)

that comes **before** 21.
22   (20)

that comes **before** 84.
(83)   85

Before you take the Self Test, study what you have read and done. The Self Test will check what you remember.

Section 2 | **13**

PAGE 14: SELF TEST 2

MATERIALS NEEDED

- pencil
- chart of numbers to 99

Concept:

number order to 99

Teacher Goal:

To teach the children to learn to check their progress periodically.

Teaching Page 14:

Turn to page 14. Read the directions to the children. Have the children repeat them after you while running their fingers under the sentence being read. Be sure the children understand what they are to do. If necessary, students may use the chart of numbers. Let the children complete the page. You may repeat the directions but give no other help.

Do not have the children check their own work. Check it as soon as you can, and go over it with each child. Show him where he did well and where he needs extra help. If necessary, take time to review before going to the next section.

NUMBER ORDER, ADD/SUBTRACT | Unit 1

SELF TEST 2

Each answer = 1 point

Put a circle around each number bigger than 50.
28   (55)   50   10   (82)   (76)

Put a circle around each number smaller than 50.
(17)   (45)   50   59   62   (39)

Write the missing number on each line.
27, 28, _29_ , _30_ , _31_ , 32, 33, _34_ , _35_ , 36
85, _86_ , 87, _88_ , 89, _90_ , 91, _92_ , 93, _94_

Put a circle around the number ...

that comes **after** 44.	that comes **before** 75.
43 (45)	76 (74)
that comes **after** 88.	that comes **before** 58.
(89) 87	59 (57)

Write the number in the box.

How many?   12

How many?   9

18
22

✔ Teacher Check

_____
Initial   Date

My Score

14 | Section 2

---

3.   ADD AND SUBTRACT TO 9

PAGE 15: ADD TO 9

MATERIALS NEEDED

• pencils
• paper
• objects for counting

Concept:

adding to 9

Teacher Goal:

To teach the children to add to 9 by using sets.

Teaching Page 15:

Review the meaning of the word *set* introduced on page 6. Place a set of *five* objects and a set of *two* objects in front of the students. Ask the students how many objects there are altogether. (seven) Place sets of *one* and *four* objects, *six* and *three* objects, and *three* and *five* objects in front of the students. Ask how many there are altogether. (five, nine, eight)

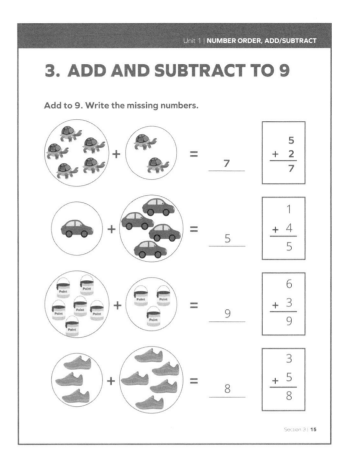

Turn to page 15. Read the instructions aloud to the students. Point to the first two circles. Ask the students what the circles represent. (two sets) Identify the + sign and the = sign. Have the students say aloud, "Five plus two equals." Ask them to point to the answer on the line. Then have them say aloud, "Five plus two equals seven." Call attention to the number fact in the box. Ask the students to find the *plus* or *add* sign. (+) Tell them the line drawn under the number fact tells them to find an answer. Have them point to the answer (7) under the line and then read the number fact aloud, "Five plus two equals seven." Proceed in the same manner to complete the page. Have the students use their objects for counting to illustrate each fact. The students should write the answers on the blank lines and then fill in the missing numbers in the number facts. Be sure the students read each number fact as it is completed.

PAGE 16: ADD TO 9

MATERIALS NEEDED

- pencils
- paper
- objects for counting
- addition fact cards on 2-inch by 3-inch cardboard using the facts as shown on page 16 —two sets may be made (one with and one without the answer on the back of the card)
- number symbol cards from page 12

Write the number in the □ .

$\begin{array}{r}1\\+\,0\\\hline 1\end{array}$	$\begin{array}{r}1\\+\,1\\\hline 2\end{array}$	$\begin{array}{r}1\\+\,2\\\hline 3\end{array}$	$\begin{array}{r}1\\+\,3\\\hline 4\end{array}$	$\begin{array}{r}1\\+\,4\\\hline 5\end{array}$	$\begin{array}{r}1\\+\,5\\\hline 6\end{array}$	$\begin{array}{r}1\\+\,6\\\hline 7\end{array}$	$\begin{array}{r}1\\+\,7\\\hline 8\end{array}$	$\begin{array}{r}1\\+\,8\\\hline 9\end{array}$
$\begin{array}{r}2\\+\,0\\\hline 2\end{array}$	$\begin{array}{r}2\\+\,1\\\hline 3\end{array}$	$\begin{array}{r}2\\+\,2\\\hline 4\end{array}$	$\begin{array}{r}2\\+\,3\\\hline 5\end{array}$	$\begin{array}{r}2\\+\,4\\\hline 6\end{array}$	$\begin{array}{r}2\\+\,5\\\hline 7\end{array}$	$\begin{array}{r}2\\+\,6\\\hline 8\end{array}$	$\begin{array}{r}2\\+\,7\\\hline 9\end{array}$	$\begin{array}{r}3\\+\,0\\\hline 3\end{array}$
$\begin{array}{r}3\\+\,1\\\hline 4\end{array}$	$\begin{array}{r}3\\+\,2\\\hline 5\end{array}$	$\begin{array}{r}3\\+\,3\\\hline 6\end{array}$	$\begin{array}{r}3\\+\,4\\\hline 7\end{array}$	$\begin{array}{r}3\\+\,5\\\hline 8\end{array}$	$\begin{array}{r}3\\+\,6\\\hline 9\end{array}$	$\begin{array}{r}4\\+\,0\\\hline 4\end{array}$	$\begin{array}{r}4\\+\,1\\\hline 5\end{array}$	$\begin{array}{r}4\\+\,2\\\hline 6\end{array}$
$\begin{array}{r}4\\+\,3\\\hline 7\end{array}$	$\begin{array}{r}4\\+\,4\\\hline 8\end{array}$	$\begin{array}{r}4\\+\,5\\\hline 9\end{array}$	$\begin{array}{r}5\\+\,0\\\hline 5\end{array}$	$\begin{array}{r}5\\+\,1\\\hline 6\end{array}$	$\begin{array}{r}5\\+\,2\\\hline 7\end{array}$	$\begin{array}{r}5\\+\,3\\\hline 8\end{array}$	$\begin{array}{r}5\\+\,4\\\hline 9\end{array}$	$\begin{array}{r}6\\+\,0\\\hline 6\end{array}$
$\begin{array}{r}6\\+\,1\\\hline 7\end{array}$	$\begin{array}{r}6\\+\,2\\\hline 8\end{array}$	$\begin{array}{r}6\\+\,3\\\hline 9\end{array}$	$\begin{array}{r}7\\+\,0\\\hline 7\end{array}$	$\begin{array}{r}7\\+\,1\\\hline 8\end{array}$	$\begin{array}{r}7\\+\,2\\\hline 9\end{array}$	$\begin{array}{r}8\\+\,0\\\hline 8\end{array}$	$\begin{array}{r}8\\+\,1\\\hline 9\end{array}$	$\begin{array}{r}9\\+\,0\\\hline 9\end{array}$

16 | Section 3

Concept:

adding to 9

Teacher Goal:

To teach the children to add to 9 by writing number facts.

Teaching Page 16:

Review the meaning of *zero* introduced on page 3. Place a set of *eight* objects in front of the students. Point to an area next to the set and tell the students that the area represents a set of *zero*. Ask the students how many objects they have altogether. (eight) Have them say aloud, "Eight plus zero equals eight." Do this with several sets of numbers so that the students grasp the idea of adding *zero* or *nothing* to a set.

Turn to page 16. Point to several of the addition number facts on the page. Ask the students to identify the *plus* or *add* sign and the line drawn under the fact. Have the students say aloud, "One plus six equals" or "Four plus three equals." Read the directions at the top of the page. Students who are able may complete the page independently. Students who need the support may use objects for counting. Follow this method. Have the students make sets to represent each number in the number fact. Then ask how many there are altogether. Students should begin to be able to see or visualize the sets in their minds. This is the first step in committing the number facts to memory.

When page 16 is complete, play a game of concentration with the students using the number symbol cards from page 12 and the number fact cards from page 16. Select twelve cards—six fact cards and the six number symbol cards that are the answers to the fact cards that you selected. Place the cards face down in three rows, four across. Have the students match the fact cards and the answer cards. The student with the most matches wins the game.

PAGE 17: ADD TO 9

MATERIALS NEEDED

- pencils
- objects for counting
- addition fact cards

Concept:

adding to 9

Teacher Goal:

To teach the children to add to 9 by writing number facts.

Teaching Page 17:

Turn to page 17. Before the students begin working the problems, ask them to read each number fact in the first row, "Two plus seven equals," "Three plus one equals," and so on. Have them use the objects for counting to illustrate each fact in the first row. Tell them to make a set of *two* and a set of *seven*. Tell them to say the fact that they have illustrated, "Two plus seven equals nine." Then have them write the *9* in the box below the line. Use this method to complete each of the facts in the first row. Students who know their facts may complete the page independently. Those who do not know their facts should continue using the objects for counting to make sets. After the students have made the sets, have them close their eyes and picture the sets. Ask them to give you the answer orally before writing the answer in the box.

Drill the students using the addition fact cards. Follow the same order as the facts shown on page 16. Identify the facts with which the students are having difficulty and set aside those fact cards. Work with the students on a regular basis using fact cards and objects for counting until the addition facts through *9* are mastered.

Write the number in the ☐.

2 +7 = 9	3 +1 = 4	4 +4 = 8	2 +4 = 6	1 +3 = 4	6 +3 = 9
2 +3 = 5	2 +1 = 3	5 +2 = 7	6 +1 = 7	2 +6 = 8	7 +2 = 9
1 +6 = 7	5 +1 = 6	1 +4 = 5	3 +2 = 5	3 +5 = 8	4 +5 = 9
4 +3 = 7	1 +5 = 6	7 +1 = 8	1 +2 = 3	3 +4 = 7	3 +5 = 8
8 +1 = 9	3 +3 = 6	5 +3 = 8	6 +3 = 9	4 +2 = 6	5 +4 = 9
5 +2 = 7	1 +8 = 9	1 +1 = 2	6 +2 = 8	4 +1 = 5	2 +2 = 4

Section 3 | **17**

PAGE 18: SUBTRACT TO 9

MATERIALS NEEDED

- pencils
- objects for counting

Concept:

subtracting to 9

Teacher Goal:

To teach the children to subtract to 9 by using sets.

Teaching Page 18:

Tell the students that in their last three lessons, they have been talking about addition. Give the students *one* object and tell them to add the object to a set of *six*. Ask "how many" they have now. (seven) Ask them if addition makes their answer *bigger* or *smaller*. (bigger) Continue illustrating this for students who do not give the correct answer. (bigger) Tell the children that today they will learn to make the sets *smaller*. Ask them if they know what this is called. (subtraction or take away) Place the set of *six* in front of the students again, and ask the students to subtract or take away *one* object from the set. Ask how many are left. (five)

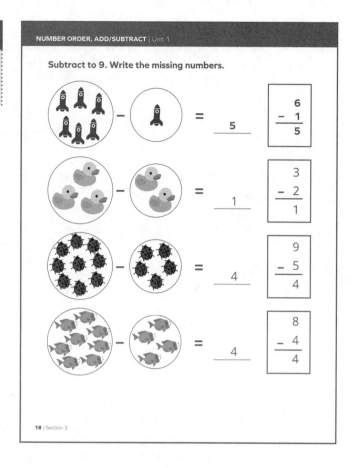

Turn to page 18. Read the instructions aloud to the students. Point to the first two circles. Ask the students what the circles represent. (two sets) Identify the − sign and the = sign. Have the students say aloud, "Six take away one equals." Ask them to point to the answer on the line. Explain to the children that we use the word *minus* when reading a subtraction problem. Have them say aloud, "Six minus one equals five." Point to the number fact in the box. Ask the students to find the *minus* or *subtraction* sign. (−) Ask them what the line drawn under the number fact means. (to find the answer) Have them point to the answer (5) under the line and then read the number fact aloud, "Six minus one equals five."

Proceed in the same manner to complete the page. Have the students use their objects for counting to illustrate each fact. The students should write the answers on the blank lines and then fill in the missing numbers in the number facts. Be sure the students read each number fact as it is completed.

PAGE 21: SELF TEST 3

MATERIALS NEEDED

- pencils
- objects for counting
- addition and subtraction fact cards

Concept:

adding and subtracting to 9

Teacher Goal:

To teach the children to learn to check their progress periodically.

Teaching Page 21:

Turn to page 21. Read the directions to the children. Have the children repeat them after you while running their fingers under the sentence being read. Be sure the children understand what they are to do. If necessary, students may use the objects for counting. Let the children complete the page. You may repeat the directions but give no other help.

Do not have the children check their own work. Check it as soon as you can, and go over it with each child. Show him where he did well and where he needs extra help. Continue reviewing the addition and subtraction fact cards as necessary.

Unit 1 | **NUMBER ORDER, ADD/SUBTRACT**

SELF TEST 3

Each answer = 1 point

Write the number in the ☐.

2	4	7	0	6	3	5	2	1
$+7$	$+4$	$+1$	$+3$	$+2$	$+6$	$+4$	$+5$	$+3$
9	8	8	3	8	9	9	7	4

4	3	5	1	0	8	2	4	2
$+1$	$+2$	$+3$	$+5$	$+5$	$+1$	$+4$	$+3$	$+1$
5	5	8	6	5	9	6	7	3

5	7	9	6	5	8	6	9	9
-1	-6	-2	-1	-2	-6	-3	-1	-7
4	1	7	5	3	2	3	8	2

8	9	6	9	7	4	9	8	3
-4	-3	-4	-8	-2	-2	-5	-2	-1
4	6	2	1	5	2	4	6	2

$\frac{29}{36}$

✔ Teacher Check

Initial   Date

My Score

Section 3 | **21**

4. MEASUREMENTS AND SHAPES

PAGE 22: MEASUREMENTS

MATERIALS NEEDED

- pencils
- objects of different sizes
- empty containers
- scale (any type of scale that measures weight)

Concept:

measurements—big and little

Teacher Goal:

To teach the children to measure by recognizing big and little.

Teaching Page 22:

Write the word *measure* on the board and have the students say the word aloud. Begin a discussion with the students about the meaning of the word *measure*. "Why do we want or need to measure things?" "What kinds of things do we measure?" "What method do we use to measure?" Use the objects to compare sizes such as *big* and *little*, *long* and *short*.

Fill the containers with varying amounts of water and compare *more than* and *less than*, *empty* and *full*. Have the students measure their weight on a scale. Compare different weights. Emphasize that "to measure" means many things.

Turn to page 22. Ask the children to identify the pictures on the page. Explain to them that today they will measure by identifying *big* and *little* objects. Read the directions to the children and have them repeat them aloud as they follow the words. Be sure they understand there are two sets of directions on the page. Allow them to circle the answers independently.

Complete the page by discussing with the children the relationship between the objects in the *big* and *little* pictures. (mother turtle, baby turtle; one boat appears close, the other appears far away; big socks belong to dad, little socks belong to child)

PAGE 23: MEASUREMENTS

MATERIALS NEEDED

• pencils
• objects for counting

Concept:

measurements—greater than, less than

Teacher Goal:

To teach the children to measure by recognizing greater than and less than.

Teaching Page 23:

Write the word *measure* on the board. Ask the children what they measured on page 22. (big and little) Introduce the expressions *greater than* and *less than*. Place sets of three and six objects in front of the students. Ask them if three is *greater than* or *less than* six. Ask them if six is *greater than* or *less than* three. Continue doing this with sets of different sizes until the students understand the expressions of *greater than* and *less than*.

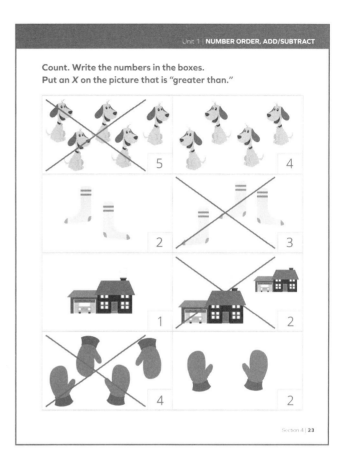

Turn to page 23. Have the students identify the pictures on the page. Explain to the students that there are two steps to each problem. Read the directions aloud and ask the students to tell the two things they must do to complete the first problem. (count the dogs—decide which number is greater) Have the students complete the first problem, then let them finish the page independently.

PAGE 24: MEASUREMENTS

MATERIALS NEEDED

- pencils
- objects that illustrate size, long and short
- boxes and/or containers

Concept:

measurements—long and short

Teacher Goal:

To teach the children to measure by recognizing long and short.

Teaching Page 24:

Write the word *measure* on the board. Ask the children what they have measured in their exercises on pages 22 and 23. (big and little, greater than, less than) Use the objects and containers to have a discussion. Ask the students to estimate answers. "Is this object too big to fit in that box?" "Will this amount of water fit in that container?" "Can we use this piece of paper to wrap that box?" Have the students use the expressions they have learned to give their answers. (too big, too little, greater than, less than, too long, too short) Allow them to illustrate their answers as much as possible.

Turn to page 24. Ask the students to identify the pictures. Have them read the directions aloud with you. Allow them to complete the page independently.

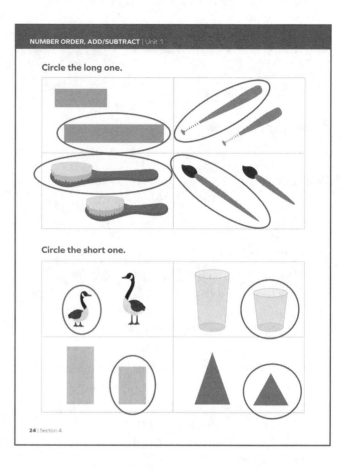

PAGE 25: MEASUREMENTS

MATERIALS NEEDED

• pencils
• cardboard
• scissors
• paste/glue
• crayons
• ruler to be made in this exercise (keep for later lessons)

Concept:

measurements—inches

Teacher Goal:

To teach the children to measure how long by using a ruler.

Teaching Page 25:

Turn to page 25. Direct the students' attention to the bottom of the page. Tell the children that today they are going to make a ruler. Write the word *ruler* on the board and have the students say it aloud. Allow the students to use the scissors to cut out the ruler at the bottom of the page. Paste the ruler to a piece of cardboard and then cut the cardboard to the same size as the ruler. (The teacher may need to help with this part of the cutting.) Tell the students to point to each number on the ruler and say it aloud. Explain to the students that each one of the numbers represents one inch. Have the students point to the numbers again and say, "One inch, two inches, three inches, four inches, five inches, six inches."

Read the directions at the top of page 25. Tell the students that they should use their rulers to measure the parts of the house. Read each direction aloud to the students and explain how to place the ruler to find the measurement. Have them write the measurement on the blank line. When the measurements are completed, read the story aloud. Ask the children what is missing from the house. (window) Allow them to color the house adding Matt the cat in the window, a drawing of themselves, or anything they would like to add to the picture. Talk to the children as they color to find out whether they can identify the colors as introduced on page 9.

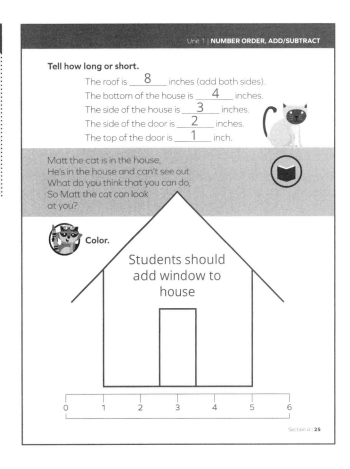

Unit 1 | **NUMBER ORDER, ADD/SUBTRACT**

Tell how long or short.

The roof is ____8____ inches (add both sides).
The bottom of the house is ____4____ inches.
The side of the house is ____3____ inches.
The side of the door is ____2____ inches.
The top of the door is ____1____ inch.

Matt the cat is in the house,
He's in the house and can't see out.
What do you think that you can do,
So Matt the cat can look at you?

Color.

Students should add window to house

0   1   2   3   4   5   6

Section 4 | **25**

PAGE 26: SHAPES

MATERIALS NEEDED

- pencils
- scissors
- colored paper
- paste/glue
- crayons

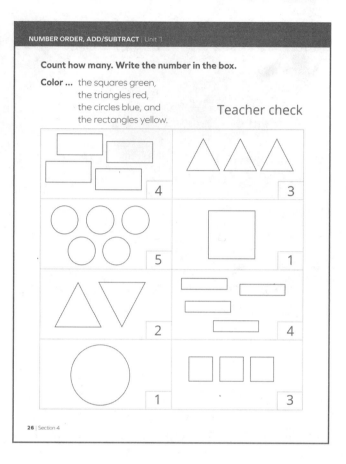

Concept:

shapes—circles, squares, triangles, rectangles

Teacher Goal:

To teach the children to recognize the shapes of circle, square, triangle, and rectangle.

Teaching Page 26:

Use the scissors and paper to cut out a variety of circles, squares, triangles, and rectangles. Ask the students to identify the shapes as they are being cut out. Allow the students to select some shapes and cut them out of the colored paper. Have them group the cutouts first by shapes, then by color. Ask which groups are *greater than* or *less than*. Have the students arrange the shapes in order from *biggest* to *smallest*. Set the shapes aside.

Turn to page 26. Ask the students to identify each group of shapes on the page. Read the instructions aloud. Tell the students that these are two-step problems and ask them to tell the steps. (count how many, color the shapes) Allow the students to write "how many." Then, repeat the coloring directions to them as necessary.

When page 26 is completed, some students may enjoy making a collage using the shapes that were cut out and set aside.

PAGE 27: ORAL PROBLEMS

MATERIALS NEEDED

• pencils

Concept:

solving oral problems

Teacher Goal:

To teach the children to listen to and solve oral problems in addition and subtraction to improve concentration.

Teaching Page 27:

Turn to page 27. Ask the students how many pictures there are on the page. (eight) Tell them that you are going to read them a story about each picture and then ask a question about the story. Explain to them that they should write the answer to the question on the blank line. Have the students point to the blank line that goes with each picture. Tell them to listen carefully.

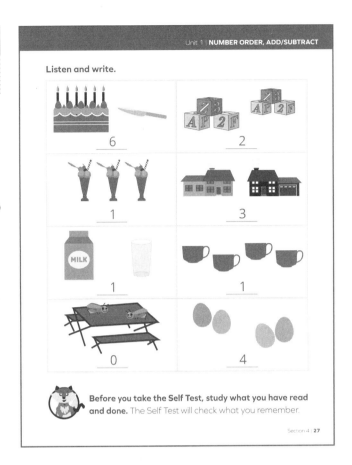

1. It is Jane's birthday. Count the candles on the cake. How old do you think Jane is?

2. Jeremy wants to build a tower with his blocks. He wants to use eight blocks to build his tower. How many more blocks does Jeremy need to build his tower?

3. Billy went to the ice cream store and ordered three sodas. He gave one to his friend Tom and one to his friend Jack. How many sodas did Billy have left?

4. Mary's best friend Julie lives in the house next door. Mary went to Julie's house to play on Monday, Tuesday, and Thursday. How many times did Mary play at Julie's house?

5. Andy and Ben have been playing outside and are very thirsty. They have enough milk to share but not enough glasses. How many more glasses do they need?

6. Laura's mother wants to hang her coffee mugs on hooks in the cupboard. She has three hooks. How many more does she need to hang the cups?

7. Beth's brother loves to play with bugs. He found two bugs on the picnic table and put them in his pocket. Now how many bugs are on the table?

8. Don's mother wanted to bake a cake. She had two eggs and she borrowed two eggs from her neighbor. How many eggs did she use altogether?

The students should prepare for the Self Test. Ask the students to look over and read the Self Test but they should not write the answers to any questions. After looking over the Self Test the students should go to the beginning of the unit and reread the text and review the answers to the activities up to the Self Test.

The students are to complete the Self Test the next school day. This should be done under regular test conditions without allowing the students to look back. A good idea is to clip the pages together before the test.

PAGE 28: SELF TEST 4

MATERIALS NEEDED

- pencils
- chart of numbers to 99

Concepts:

measurements, shapes, oral directions

Teacher Goal:

To teach the children to learn to check their progress periodically.

Teaching Page 28:

Turn to page 28. Read the directions to the children. Have them complete each exercise as you read the directions. The last three exercises are oral story problems.

Listen and write.

> Katie's mother baked one cherry pie, one apple pie, and one blueberry pie. How many pies did she bake altogether?

> Donna has six colors in her color dish. She wants to have a new brush to use with each color. How many more brushes does Donna need?

> Paul has a little Bible just like his parents' big Bible. Paul loaned his Bible to his friend. How many Bibles are left in Paul's family?

Do not have the children check their own work. Check it as soon as you can, and go over it with each child. Show him where he did well and where he needs extra help. Continue reviewing the addition and subtraction fact cards as necessary.

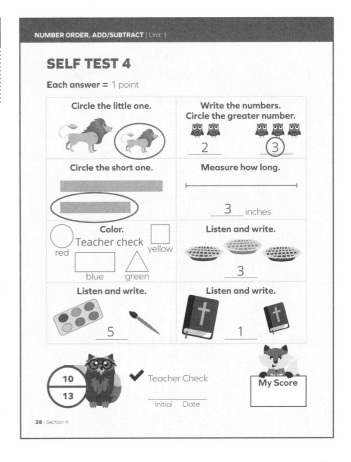

NUMBER ORDER, ADD/SUBTRACT | Unit 1

SELF TEST 4

Each answer = 1 point

Circle the little one.

Write the numbers. Circle the greater number.
2   3

Circle the short one.

Measure how long.
3 inches

Color. Teacher check
red   yellow
blue   green

Listen and write.
3

Listen and write.
5

Listen and write.
1

10
13
Teacher Check
Initial   Date

My Score

28 | Section 4

5. NUMBER ORDER

PAGE 29: LISTENING SKILLS

MATERIALS NEEDED

- pencils

Concept:

listening and writing

Teacher Goal:

To teach the children to apply auditory skills to write numbers to 99.

Teaching Page 29:

Turn to page 29. Read the directions with the children and have them identify the animals on the page. Tell them that you will say the name of each animal and that they should write their answers in the box that has that animal. Tell them to listen carefully.

Find the caterpillar. There are eight lines. Start with the number *5*. Write numbers in order to *12* on the lines.

Find the turtle. There are eight lines. Start with the number *28*. Write numbers in order to *35* on the lines.

Find the duck. There are eight lines. Write the *answers* to these eight facts.

1 + 2 =	4 + 3 =	5 + 0 =	7 + 1 =
4 – 3 =	6 – 2 =	8 – 4 =	3 – 2 =

Find the donkey. There are eight lines. Write the number that comes after ...

5	18	22	37
49	52	60	84

Find the pig. There are eight lines. Write the number that comes before ...

7	14	22	38
40	73	87	100

Unit 1 | **NUMBER ORDER, ADD/SUBTRACT**

5. NUMBER ORDER

Listen and write.

5	6	7	8	9	10	11	12
28	29	30	31	32	33	34	35

3	7	5	8
1	4	4	1

6	19	23	38
50	53	61	85

6	13	21	37
39	72	86	99

Section 5 | **29**

PAGES 30 & 31: NUMBERS TO 99

MATERIALS NEEDED

• pencils

Concept:

number relationships

Teacher Goal:

To teach the children to tell "How many?"

Teaching Pages 30 and 31:

Turn to page 30. Ask the students to point to the words "How many?" and read them aloud. Have the children find the picture of the dogs. Tell them to count the dogs and write the number in the box. Instruct them to say the words "How many?" as they continue down the page, writing an answer in each box.

When the children have completed the exercise draw their attention to the number line at the top of page 30. Turn to page 31. Tell the students they may use the number line to answer the questions on page 31. In the first exercise, they are to write the numbers they have written in the boxes on page 30 in number order. In the second exercise, they should write the numbers to 19 that are missing from their list. In the third exercise, *they should refer to the count in the boxes on page 30* and *add to* or *subtract from* each number.

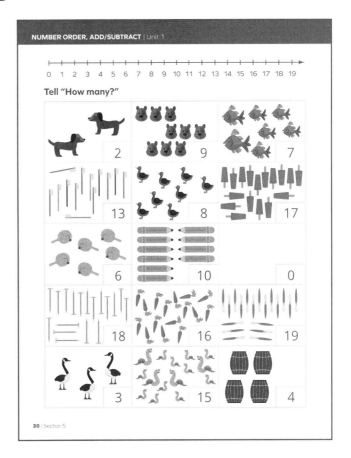

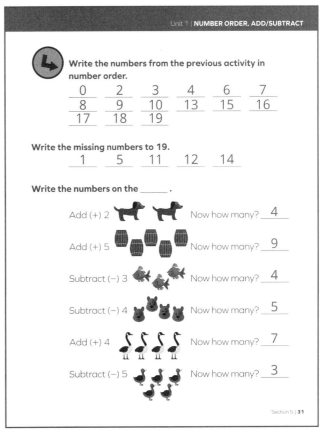

PAGE 32: FLAT SHAPES AND COLORS

MATERIALS NEEDED

- pencils
- crayons
- scissors
- paper

Concepts:

flat shapes, colors

Teacher Goals:

To teach the children to recognize flat shapes and to identify colors.

Teaching Page 32:

Turn to page 32. Ask the students the name of the shapes on page 32 (flat shapes) and ask them to identify each one—circle, triangle, square, rectangle. Read the directions with the children and have them complete the page. Spend some time when the page is completed reviewing shapes and colors with the students. Have them use scissors, paper, and crayons to cut out and color their own shapes.

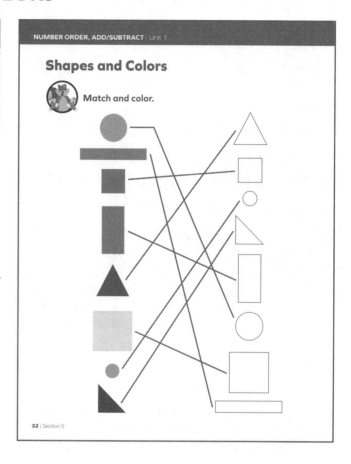

NUMBER ORDER, ADD/SUBTRACT | Unit 1

Shapes and Colors

Match and color.

32 | Section 5

PAGE 33: NUMBER FACTS

MATERIALS NEEDED

• pencils
• crayons

Concept:

addition and subtraction facts

Teacher Goals:

To teach the children to practice addition and subtraction facts and to identify "greater than" and "less than" numbers.

Teaching Page 33:

Turn to page 33. Tell the children that they will be practicing their number facts. Read the directions on the page with them. Be sure they are reading the addition (+) and subtraction (–) signs correctly. When they have finished each set, have them compare the answers and, using a crayon of their own choosing, circle the number that is *greater than* or *less than*.

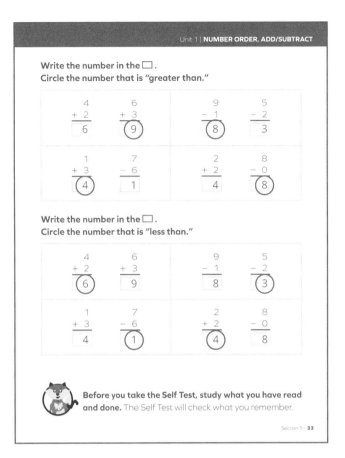

The students should prepare for the Self Test. Ask the students to look over and read the Self Test but they should not write the answers to any questions. After looking over the Self Test the students should go to the beginning of the unit and reread the text and review the answers to the activities up to the Self Test.

The students are to complete the Self Test the next school day. This should be done under regular test conditions without allowing the students to look back. A good idea is to clip the pages together before the test.

PAGE 34: SELF TEST 5

MATERIALS NEEDED

• pencils

Concept:

number order

Teacher Goal:

To teach the children to learn to check their progress periodically.

Teaching Page 34:

Turn to page 34. Read the directions to the children. Have them complete each exercise as you read the directions.

Dictate the questions to *Listen and write*.

Find the caterpillar. There are six lines. Start with the number 42 and write the numbers to 47.

Find the duck. There are four lines. Write the numbers after ...

> 12   43   79   94

Find the donkey. There are four lines. Write the numbers before ...

> 20   37   54   70

Find the pig. There are four shapes. Write the number 18 in the triangle, 22 in the circle, 43 in the square, and 56 in the rectangle.

Have the children complete the facts.

Do not have the children check their own work. Check it as soon as you can, and go over it with each child. Show him where he did well and where he needs extra help.

The students should prepare for the LIFEPAC Test. The students should go to the beginning of the unit and reread the text and review the answers to the activities for the entire unit. Ask the students questions to check their understanding of the unit.

The students are to complete the LIFEPAC Test the next school day. This should be done under regular test conditions without allowing the students to look at the unit.

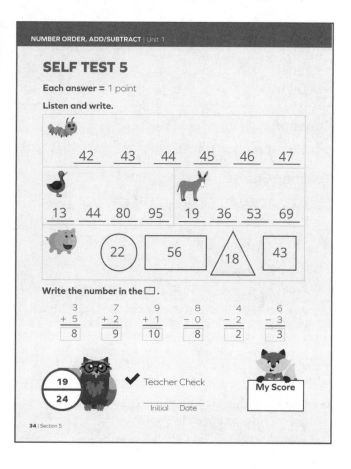

NUMBER ORDER, ADD/SUBTRACT | Unit 1

SELF TEST 5

Each answer = 1 point

Listen and write.

42   43   44   45   46   47

13   44   80   95   19   36   53   69

22   56   18   43

Write the number in the ☐.

3	7	9	8	4	6
+ 5	+ 2	+ 1	− 0	− 2	− 3
8	9	10	8	2	3

19 / 24

✔ Teacher Check

Initial   Date

My Score

34 | Section 5

LIFEPAC TEST 101

MATERIALS NEEDED

- pencils
- crayons

Concepts:

count numbers to 99, add and subtract to 9, recognize greater than and less than, measure to the inch, recognize flat shapes and colors, solve oral story problems

Teacher Goal:

To teach the children to learn to check their own progress periodically.

Teaching the LIFEPAC Test:

Administer the test in at least two sessions.

Read all of the directions on each page as the children prepare to do it. Be sure that they understand what they are being asked to do.

LIFEPAC Test page 4

Listen and write.

(2 points each: 1 for correct shape, 1 for correct color)

Put a ...

1 on the square, color it blue.

2 on the triangle, color it pink.

3 on the circle, color it green.

4 on the rectangle, color it orange.

(2 points: 1 for addition, 1 for correct answer)

Lucie found 6 crayons in one box and 3 crayons in another box. How many crayons did she find altogether?

Give no help except with directions.

Go over each page with the child as soon as possible after you check it so that he can see where he did well and where he needs more work.

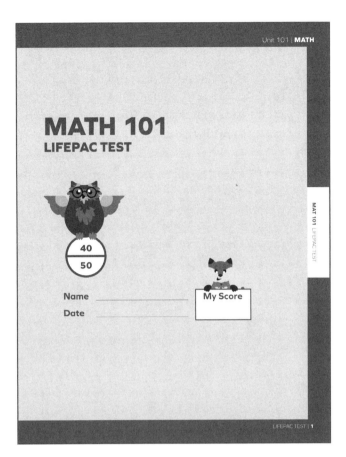

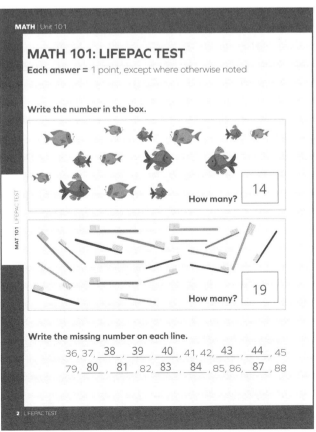

Evaluate the tests and review areas where the children have done poorly. Review the pages and activities that stress the concepts tested.

If necessary, when the children have reviewed sufficiently, administer the Alternate LIFEPAC test. Follow the same procedures as used for the LIFEPAC Test.

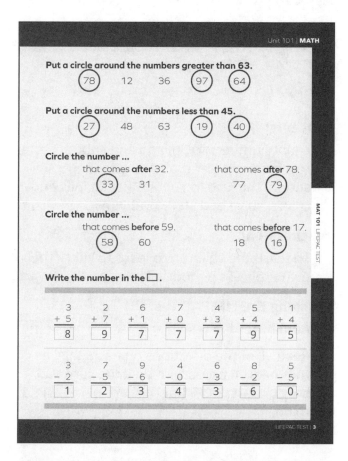

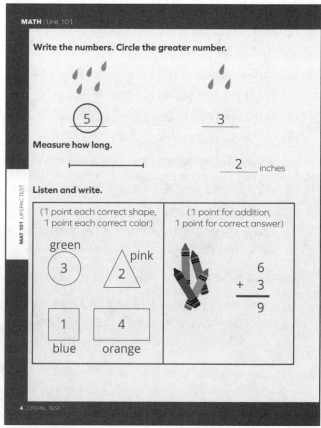

ALTERNATE LIFEPAC TEST 101

MATERIALS NEEDED

- pencils
- crayons

Concepts:

count numbers to 99, add and subtract to 9, recognize greater than and less than, measure to the inch, recognize flat shapes and colors, solve oral story problems

Teacher Goal:

To teach the children to learn to check their own progress periodically.

Teaching the Alternate LIFEPAC Test:

Administer the test in at least two sessions.

Read all of the directions on each page as the children prepare to do it. Be sure that they understand what they are being asked to do.

Alternate LIFEPAC Test page 4

Listen and write.

(2 points each: 1 for correct shape, 1 for correct color)

Put a ...

1 on the circle, color it red.

2 on the square, color it brown.

3 on the rectangle, color it black.

4 on the triangle, color it purple.

(2 points: 1 for addition, 1 for correct answer)

Paul counted 6 birds in his yard on Monday and 3 birds in his yard on Tuesday. How many birds did he count altogether?

Give no help except with directions.

Go over each page with the child as soon as possible after you check it so that he can see where he did well and where he needs more work.

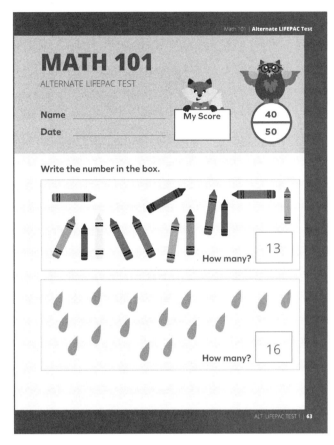

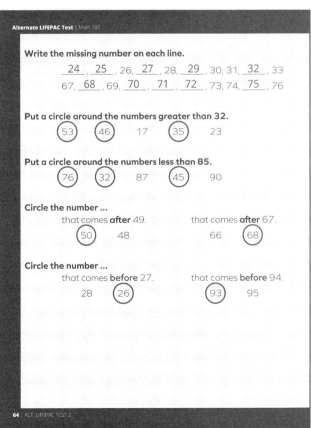

Evaluate the tests and review areas where the children have done poorly. Review the pages and activities that stress the concepts tested.

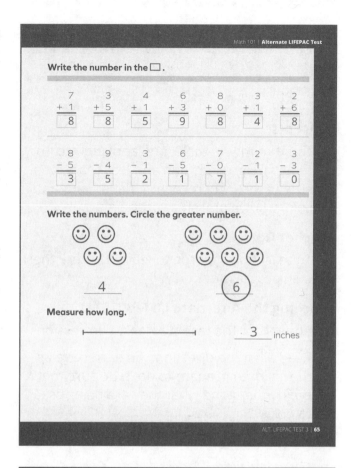

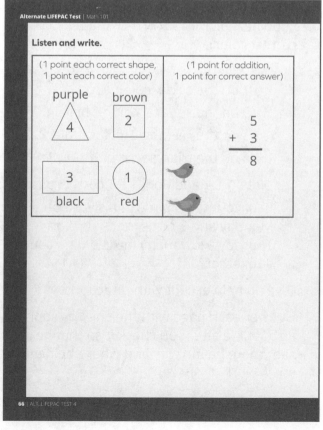

MATH 101

ALTERNATE LIFEPAC TEST

Name _____

Date _____

My Score

40

50

Write the number in the box.

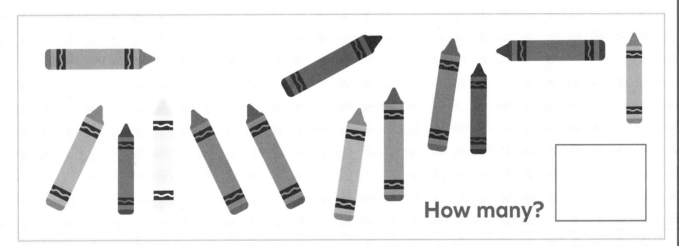

How many?

How many?

Write the missing number on each line.

_____ , _____ , 26, _____ , 28, _____ , 30, 31, _____ , 33

67, _____ , 69, _____ , _____ , _____ , 73, 74, _____ , 76

Put a circle around the numbers greater than 32.

53   46   17   35   23

Put a circle around the numbers less than 85.

76   32   87   45   90

Circle the number ...

that comes **after** 49.   that comes **after** 67.

50   48   66   68

Circle the number ...

that comes **before** 27.   that comes **before** 94.

28   26   93   95

Write the number in the ☐ .

7	3	4	6	8	3	2
+ 1	+ 5	+ 1	+ 3	+ 0	+ 1	+ 6
☐	☐	☐	☐	☐	☐	☐

8	9	3	6	7	2	3
− 5	− 4	− 1	− 5	− 0	− 1	− 3
☐	☐	☐	☐	☐	☐	☐

Write the numbers. Circle the greater number.

_____   _____

Measure how long.

├─────────────────────┤   _____ inches

Listen and write.

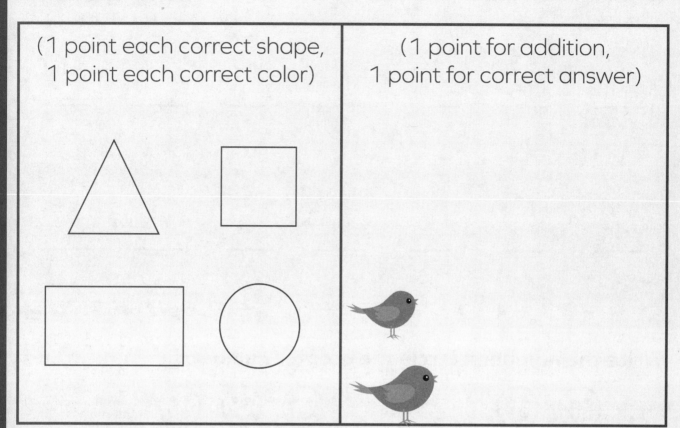

(1 point each correct shape,
1 point each correct color)

(1 point for addition,
1 point for correct answer)

MATH 102

Unit 2: Add/Subtract to 10, Shapes

ADD/SUBTRACT TO 10, SHAPES
MATH 102

Alpha Omega
PUBLICATIONS

**804 N. 2nd Ave. E.
Rock Rapids, IA 51246-1759**

Author:
Carol Bauler, B.A.

Editor:
Alan Christopherson, M.S.

Media Credits:
Page 1: © wenchiawang, iStock, Thinkstock;
24: © blueringmedia, iStock, Thinkstock.

i

PAGE 1: ADD/SUBTRACT TO 10, SHAPES

MATERIALS NEEDED

• pencils

Concepts:

purpose of LIFEPAC, objectives

Teacher Goals:

To teach the children to know what is expected of the student in the LIFEPAC and to write first and last names correctly.

Teaching Page 1:

Turn to page 1. Point to the title and read it aloud. Allow time for the children to look through the LIFEPAC. Write the word *Objectives* on the board and have the children find the word on the page. Explain that the objectives tell the things the students will be expected to do in the LIFEPAC. Read each one and have the children repeat them as they run their fingers under the sentence from left to right. Talk about the objectives so that the children will understand what they will be doing. Have each child write his name on the line.

1. ADDITION FACTS TO 10

PAGE 2: FACTS TO 10

MATERIALS NEEDED

- pencils
- paper
- objects for counting
- new fact cards (for 0 + 10, 1 + 9, 2 + 8, 3 + 7, 4 + 6, 5 + 5, 6 + 4, 7 + 3, 8 + 2, 9 +1, 10 + 0)

Concept:

add facts to 10

Teacher Goal:

To teach the children to add number facts to 10 vertically.

Teaching Page 2:

Draw a number line on the board showing the numbers *0* through *10*. Have the students count the numbers aloud. Place a set of *1* object in front of the students and have them show how many more objects they would need to count to *10*. Tell them to point to the *1* on the number line and have them count how many more to *10*. Ask the children to write the number fact they have illustrated on paper vertically (up and down). Continue this type of exercise until you have illustrated each of the new fact cards. Go through the fact cards once more.

Turn to page 2. Point to the number line at the top of the page and then read the directions with the students. Allow them to complete the page independently.

ADD/SUBTRACT TO 10, SHAPES | Unit 2

1. ADDITION FACTS TO 10

0 1 2 3 4 5 6 7 8 9 10

Write the number in the □.

1	2	3	4	6	5	1	2	7
+ 6	+ 5	+ 3	+ 5	+ 3	+ 5	+ 3	+ 8	+ 3
7	7	6	9	9	10	4	10	10
8	5	5	5	3	6	4	8	2
+ 2	+ 0	+ 1	+ 3	+ 1	+ 4	+ 0	+ 1	+ 1
10	5	6	8	4	10	4	9	3
1	9	7	4	2	2	8	5	3
+ 8	+ 0	+ 1	+ 1	+ 2	+ 7	+ 0	+ 2	+ 6
9	9	8	5	4	9	8	7	9
4	6	1	2	2	9	5	3	7
+ 2	+ 2	+ 0	+ 4	+ 0	+ 1	+ 4	+ 0	+ 0
6	8	1	6	2	10	9	3	7
3	2	1	4	7	6	1	1	3
+ 5	+ 6	+ 1	+ 4	+ 2	+ 0	+ 9	+ 4	+ 7
8	8	2	8	9	6	10	5	10
6	4	2	4	3	3	1	1	1
+ 1	+ 3	+ 3	+ 6	+ 4	+ 2	+ 2	+ 7	+ 5
7	7	5	10	7	5	3	8	6

2 | Section 1

PAGES 3 & 4: ADDITION FACTS

MATERIALS NEEDED

- pencils
- new operation symbol cards for +, −, =
- number symbol cards (LIFEPAC 101, page 12)
- addition fact cards (LIFEPAC 101, page 16; LIFEPAC 102, page 2)

Concept:

add number facts

Teacher Goal:

To teach the children to add number facts horizontally and vertically.

Teaching Pages 3 and 4:

Turn to pages 3 and 4. Read the instructions at the top of the page with the children. Have them count aloud the number of dogs in the first two sets. Point to the symbols (+, =). Ask the children to read the problem aloud. (Three dogs plus two dogs equals five dogs.) Have the children point to the two number facts. Tell them that there are two ways to write their number facts, horizontally (across) and vertically (up and down). Ask them to read the number facts again. Follow the same procedure for the blocks. Ask the children to read the problem aloud as they point to the horizontal number fact. Tell them to fill in the vertical number fact. Have them fill in the blanks for the bunnies. They should follow the same steps to complete page 4.

Have the children use the number symbol cards and the new operation symbol cards to make addition number facts. Explain how the cards can be rearranged to show the facts horizontally and vertically. Ask the children to illustrate several number facts using the cards.

Select a group of number symbol cards and addition fact cards. Play a game of concentration.

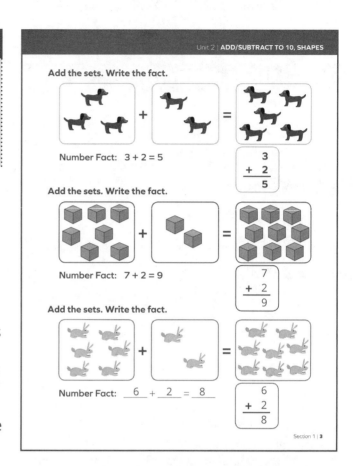

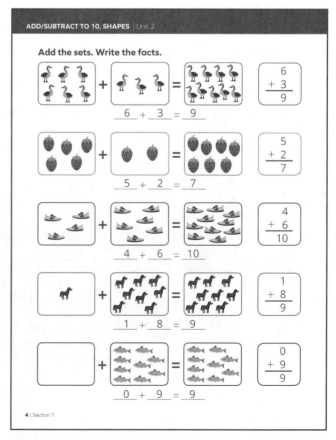

PAGE 5: ADDITION FACTS

MATERIALS NEEDED

• pencils

Concept:

add number facts horizontally

Teacher Goal:

To teach the children to add number facts to 10 horizontally.

Teaching Page 5:

Turn to page 5. Read the directions and have the children complete the number facts independently. When they have completed the page, select several facts at random and ask the students to read the facts aloud.

The students should prepare for the Self Test. Ask the students to look over and read the Self Test but they should not write the answers to any questions. After looking over the Self Test the students should go to the beginning of the unit and reread the text and review the answers to the activities up to the Self Test.

The students are to complete the Self Test the next school day. This should be done under regular test conditions without allowing the students to look back. A good idea is to clip the pages together before the test.

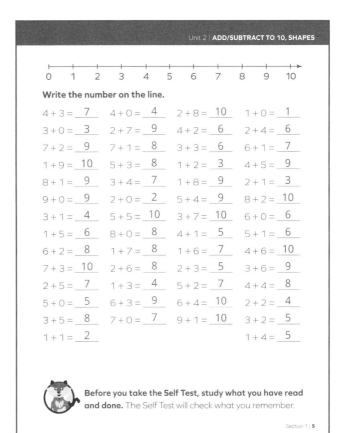

Unit 2 | **ADD/SUBTRACT TO 10, SHAPES**

Write the number on the line.

4 + 3 = 7	4 + 0 = 4	2 + 8 = 10	1 + 0 = 1
3 + 0 = 3	2 + 7 = 9	4 + 2 = 6	2 + 4 = 6
7 + 2 = 9	7 + 1 = 8	3 + 3 = 6	6 + 1 = 7
1 + 9 = 10	5 + 3 = 8	1 + 2 = 3	4 + 5 = 9
8 + 1 = 9	3 + 4 = 7	1 + 8 = 9	2 + 1 = 3
9 + 0 = 9	2 + 0 = 2	5 + 4 = 9	8 + 2 = 10
3 + 1 = 4	5 + 5 = 10	3 + 7 = 10	6 + 0 = 6
1 + 5 = 6	8 + 0 = 8	4 + 1 = 5	5 + 1 = 6
6 + 2 = 8	1 + 7 = 8	1 + 6 = 7	4 + 6 = 10
7 + 3 = 10	2 + 6 = 8	2 + 3 = 5	3 + 6 = 9
2 + 5 = 7	1 + 3 = 4	5 + 2 = 7	4 + 4 = 8
5 + 0 = 5	6 + 3 = 9	6 + 4 = 10	2 + 2 = 4
3 + 5 = 8	7 + 0 = 7	9 + 1 = 10	3 + 2 = 5
1 + 1 = 2			1 + 4 = 5

Before you take the Self Test, study what you have read and done. The Self Test will check what you remember.

Section 1 | **5**

PAGE 6: SELF TEST 1

MATERIALS NEEDED

- pencils
- objects for counting
- addition fact cards

Concept:

addition facts to 10

Teacher Goal:

To teach the children to learn to check their progress periodically.

Teaching Page 6:

Turn to page 6. Read the directions to the children. Have the children repeat them after you while running their fingers under the sentence being read. Be sure the children understand what they are to do. If necessary, students may use the objects for counting. Let the children complete the page. You may repeat the directions but give no other help.

Do not have the children check their own work. Check it as soon as you can, and go over it with each child. Show him where he did well and where he needs extra help. Continue reviewing the addition fact cards as necessary.

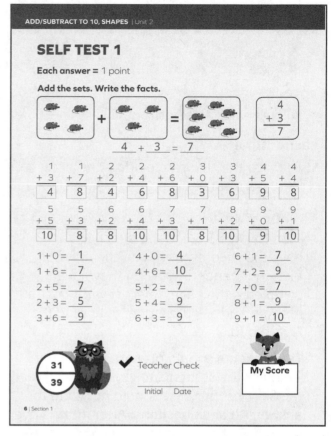

2. SUBTRACTION FACTS TO 10

PAGE 7: FACTS TO 10

MATERIALS NEEDED

- pencils
- paper
- objects for counting
- new fact cards (for 10 − 0, 10 − 1, 10 − 2, 10 − 3, 10 − 4, 10 − 5, 10 − 6, 10 − 7, 10 − 8, 10 − 9, 10 − 10)

Concept:

subtract facts to 10

Teacher Goal:

To teach the children to subtract number facts to 10 vertically.

Teaching Page 7:

Draw a number line on the board showing the numbers *0* through *10*. Have the students count the numbers aloud. Place a set of *10* objects in front of the students and ask them how many objects they would need to subtract (take away) to have *9* objects left. Tell the students to point to the *10* on the number line and count how many to *9*. Have the children write the number problem on paper vertically (up and down). Continue this exercise until you have illustrated each of the new fact cards. Go through the fact cards once more.

Turn to page 7. Point to the number line at the top of the page and then read the directions with the students. Allow them to complete the page independently.

Unit 2 | **ADD/SUBTRACT TO 10, SHAPES**

2. SUBTRACTION FACTS TO 10

0 1 2 3 4 5 6 7 8 9 10

Write the number in the ☐.

9 − 7 = 2	5 − 2 = 3	4 − 3 = 1	1 − 0 = 1	10 − 7 = 3	7 − 1 = 6	5 − 1 = 4	9 − 3 = 6	10 − 4 = 6
8 − 7 = 1	6 − 5 = 1	5 − 0 = 5	3 − 2 = 1	8 − 2 = 6	4 − 1 = 3	7 − 0 = 7	10 − 1 = 9	6 − 2 = 4
8 − 5 = 3	6 − 3 = 3	10 − 2 = 8	9 − 2 = 7	8 − 6 = 2	5 − 3 = 2	4 − 0 = 4	8 − 1 = 7	9 − 6 = 3
7 − 2 = 5	3 − 0 = 3	10 − 8 = 2	4 − 2 = 2	2 − 0 = 2	10 − 2 = 8	7 − 6 = 1	3 − 1 = 2	8 − 0 = 8
9 − 4 = 5	10 − 5 = 5	8 − 4 = 4	9 − 0 = 9	9 − 8 = 1	7 − 4 = 3	8 − 3 = 5	9 − 1 = 8	6 − 0 = 6
6 − 1 = 5	9 − 5 = 4	7 − 3 = 4	6 − 4 = 2	10 − 6 = 4	7 − 5 = 2	10 − 9 = 1	5 − 4 = 1	2 − 1 = 1

Section 2 | **7**

PAGES 8 & 9: SUBTRACTION FACTS

MATERIALS NEEDED

- pencils
- operation symbol cards for −, =
- number symbol cards
- subtraction fact cards (LIFEPAC 101, page 19; LIFEPAC 102, page 7)

Concept:

subtract number facts

Teacher Goal:

To teach the children to subtract number facts horizontally and vertically.

Teaching Pages 8 and 9:

Turn to pages 8 and 9. Read the instructions at the top of the page with the children. Have them count aloud the number of dogs in the first two sets. Point to the symbols (−, =). Ask the children to read the problem aloud. (Five dogs minus [take away] two dogs equals three dogs.) Have the children point to the two number facts. Tell them that there are two ways to write their number facts, horizontally (across) and vertically (up and down). Ask them to read the number facts again. Follow the same procedure for the blocks. Ask the children to read the problem aloud as they point to the horizontal number fact. Tell them to fill in the vertical number fact. Have them fill in the blanks for the bunnies. They should follow the same steps to complete page 9.

Have the children use the number symbol cards and the new operation symbol cards to make subtraction number facts. Explain how the cards can be rearranged to show the facts horizontally and vertically. Allow the children to illustrate several facts themselves.

Select a group of number symbol cards and addition fact cards. Play a game of concentration.

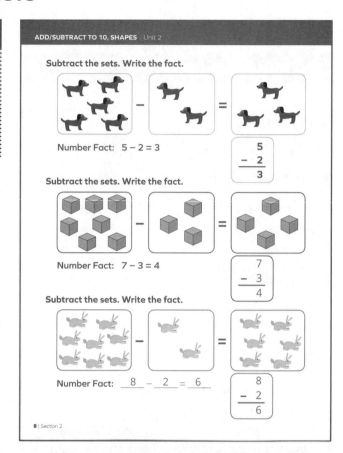

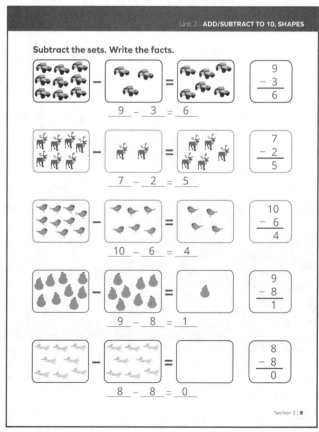

PAGE 10: SUBTRACTION FACTS

MATERIALS NEEDED

• pencils

Concept:

subtract number facts horizontally

Teacher Goal:

To teach the children to subtract number facts to 10 horizontally.

Teaching Page 10:

Turn to page 10. Read the directions and have the children complete the number facts independently. When they have completed the page, select several facts at random and ask the students to read the facts aloud.

The students should prepare for the Self Test. Ask the students to look over and read the Self Test but they should not write the answers to any questions. After looking over the Self Test the students should go to the beginning of the unit and reread the text and review the answers to the activities up to the Self Test.

The students are to complete the Self Test the next school day. This should be done under regular test conditions without allowing the students to look back. A good idea is to clip the pages together before the test.

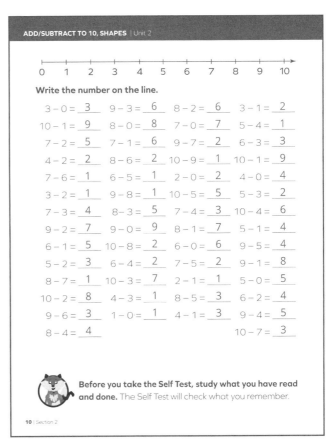

PAGE 11: SELF TEST 2

MATERIALS NEEDED

- pencils
- objects for counting
- subtraction fact cards

Concept:

subtraction facts to 10

Teacher Goal:

To teach the children to learn to check their progress periodically.

Teaching Page 11:

Turn to page 11. Read the directions to the children. Have the children repeat them after you while running their fingers under the sentence being read. Be sure the children understand what they are to do. If necessary, students may use the objects for counting. Let the children complete the page. You may repeat the directions but give no other help.

Do not have the children check their own work. Check it as soon as you can, and go over it with each child. Show him where he did well and where he needs extra help. Continue reviewing the subtraction fact cards as necessary.

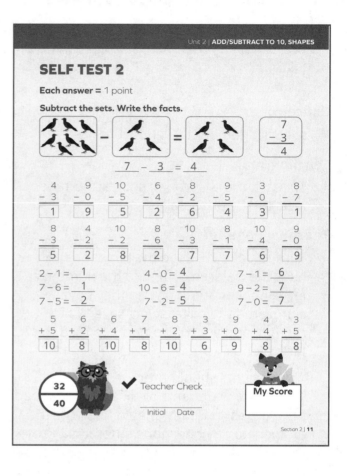

3.   ONES' AND TENS' PLACE

PAGE 12: COUNT TO 99

MATERIALS NEEDED

- pencils
- chart of numbers to 99 (LIFEPAC 101, page 7)
- objects for counting
- red crayon

Concept:

to count to 99

Teacher Goal:

To teach the children to review counting to 99.

Teaching Page 12:

Have the children use their chart of numbers to count to 99 aloud. Turn to page 12. Read the instructions with the students and together complete the first row of numbers. Allow the students to continue filling in the numbers independently.

Students should refer to the chart of numbers only when necessary. Next, have the students select ten objects for counting. Have them use their red crayons to circle the number on page 12 that represents the ten objects. (10) Ask the students to circle the number that represents two sets of ten objects. (20) Have those students who cannot answer this question make a second set of ten objects and then count the total of the two sets. Continue this type of exercise for three sets of ten objects (30), four sets (40), five sets (50), six sets (60), seven sets (70), eight sets (80), and nine sets (90). Have the students select sets in random order and say what numbers the sets represent.

ADD/SUBTRACT TO 10, **SHAPES** | Unit 2

3. ONES' AND TENS' PLACE

Write the numbers to 99.

0	1	2	3	4	5	6	7	8	9
10	11	12	13	14	15	16	17	18	19
20	21	22	23	24	25	26	27	28	29
30	31	32	33	34	35	36	37	38	39
40	41	42	43	44	45	46	47	48	49
50	51	52	53	54	55	56	57	58	59
60	61	62	63	64	65	66	67	68	69
70	71	72	73	74	75	76	77	78	79
80	81	82	83	84	85	86	87	88	89
90	91	92	93	94	95	96	97	98	99

12 | Section 3

PAGE 13: PLACE VALUE

MATERIALS NEEDED

- pencils
- objects for counting

Concept:

place value for tens and ones

Teacher Goal:

To teach the children to learn place value for tens and ones.

Teaching Page 13:

Write the number *26* on the board. Ask the students how many sets of tens there are in this number.

Introduce the expressions *tens'* place and *ones'* place to the students. Circle the number *2* and tell the students that this is called the *tens'* place because it tells us how many sets of *10* there are. Circle the *6*. Tell the students that this is called the *ones'* place because it tells us how many *ones* there are. Use objects for counting to illustrate two sets of *10* and *6* more equals *26*.

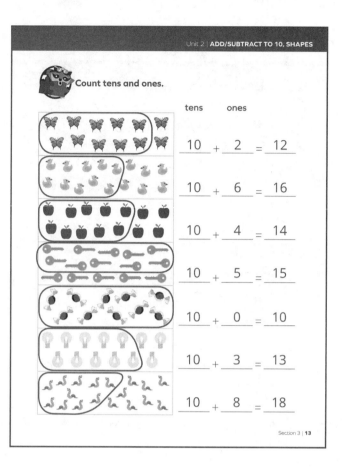

Turn to page 13. Read the directions at the top of the page. Point to the first box. Have the students circle *ten* objects and write the number *10* in the *tens'* column. Have the students count how many more objects there are (2) and write the number in the *ones'* column. Then have the students complete the last column. Monitor the students carefully as they complete the page.

PAGE 14: PLACE VALUE

MATERIALS NEEDED

• pencils
• number symbol cards
• objects for counting

Concept:

place value for tens and ones

Teacher Goal:

To teach the children to learn place value for tens and ones.

Teaching Page 14:

Place number symbol cards to represent *34* in front of the students and have them say the number aloud. Ask how many *tens* and how many *ones*. Have the students use objects for counting to make three sets of *ten* plus four *ones*. Reverse the symbols so that the number now reads *43* and have the students say the number aloud. Ask how many *tens* and how many *ones* there are now. Have the students make four sets of *ten* plus three *ones* to illustrate the number.

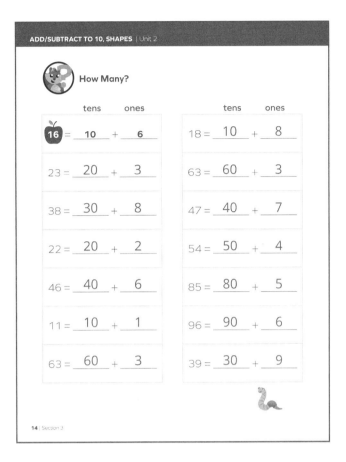

Continue this exercise with a variety of number symbols.

Turn to page 14. Point to the number *16*. Ask the students to identify the number in the *tens'* place (1) and the number in the *ones'* place (6). Tell them to point to the *10* and tell them it represents the *1* in the *tens'* place. Have them point to the *6* and tell them this represents how many *ones*. Use objects for counting to illustrate the first example. Have the students point to the numbers in *23* that represent *tens* and *ones*. Tell them to write the numbers for *tens* and *ones* on the blank lines. Use objects for counting to illustrate two sets of *ten* plus three *ones*. Those children who are able may complete the page independently; other children should continue to use objects for counting until they understand the concept.

PAGES 15 & 16: ADD TO 10

MATERIALS NEEDED

• pencils

Concept:

adding a number to 10

Teacher Goal:

To teach the children to add a number to 10.

Teaching Pages 15 and 16:

Turn to page 15 and read the title aloud. Have the class count the puppies in the pen. State, "There are ten puppies in the pen." Have the children count the puppies outside the pen. State, "Two puppies are not in the pen." Have the class read aloud the corresponding number fact. Ask which numbers are in the *ones'* place and which number is in the *tens'* place. Explain to the students that they should add the *ones'* column first and then the *tens'* column. Have the class read aloud each of the number facts at the bottom of the page. Have them add the *ones'* and *tens'* columns and read the facts again.

Turn to page 16 and read the instructions aloud. Ask the children to point to the two sets of trumpets. Have the children count the number of each set aloud. Tell the children to add the sets together. Read the corresponding number fact aloud. Have the class write the correct number in the box. Allow the children to complete the page independently.

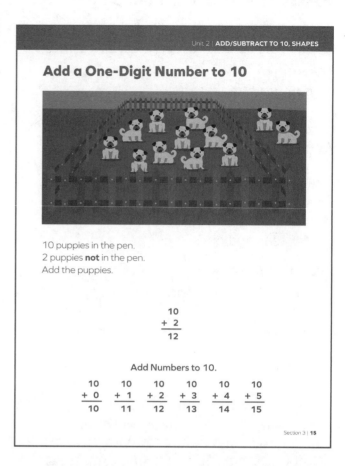

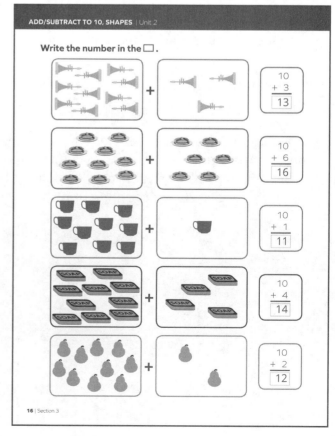

PAGE 17: ADD TO 10

MATERIALS NEEDED

- pencils
- objects for counting

Concept:

adding a number to 10

Teacher Goal:

To teach the children to add a number to 10.

Teaching Page 17:

Turn to page 17 and read the instructions aloud. Tell the children to complete the first fact by adding the numbers in the *ones'* column and then the number in the *tens'* column. Ask them to read the number fact aloud. (Ten plus two equals twelve.) Have them illustrate the fact using objects for counting. Complete each problem in the first row using these steps.

Allow the children to complete the next row independently; then, use the instructions from the first row for the third row. Let them complete the fourth row independently.

The students should prepare for the Self Test. Ask the students to look over and read the Self Test but they should not write the answers to any questions. After looking over the Self Test the students should go to the beginning of the unit and reread the text and review the answers to the activities up to the Self Test.

The students are to complete the Self Test the next school day. This should be done under regular test conditions without allowing the students to look back. A good idea is to clip the pages together before the test.

Unit 2 | **ADD/SUBTRACT TO 10, SHAPES**

Write the number in the ☐.

10	10	10	10	10
+ 2	+ 6	+ 5	+ 1	+ 0
12	16	15	11	10

10	10	10	10	10
+ 3	+ 5	+ 4	+ 9	+ 7
13	15	14	19	17

10	10	10	10	10
+ 8	+ 4	+ 3	+ 0	+ 1
18	14	13	10	11

10	10	10	10	10
+ 5	+ 7	+ 8	+ 6	+ 9
15	17	18	16	19

Before you take the Self Test, study what you have read and done. The Self Test will check what you remember.

Section 3 | **17**

PAGE 18: SELF TEST 3

MATERIALS NEEDED

- pencils
- objects for counting

Concepts:

place value for tens and ones, adding numbers to 10

Teacher Goal:

To teach the children to learn to check their progress periodically.

Teaching Page 18:

Turn to page 18. Read the directions to the children. Have the children repeat them after you while running their fingers under the sentence being read. Be sure the children understand what they are to do. If necessary, students may use the objects for counting. Let the children complete the page. You may repeat the directions but give no other help.

Do not have the children check their own work. Check it as soon as you can, and go over it with each child. Show him where he did well and where he needs extra help.

4.   NUMBER ORDER TO 99

PAGE 19: COUNT TO 99

MATERIALS NEEDED

- pencils
- chart of numbers
- crayons

Concept:

counting to 99

Teacher Goal:

To teach the children to count to 99 by drawing lines dot-to-dot.

Teaching Page 19:

Using the chart of numbers, have the students count aloud from *1* to *99*.

Turn to page 19 and read the story aloud with the children. Ask them to find the table, the cups of tea, and the teapot. Help them find the numbers *1* and *2* on the picture and explain that they will find the rhinoceros and the elephant as they connect the dots.

Allow the students to color the picture when the dot-to-dot is completed.

PAGE 20: NUMBER ORDER

MATERIALS NEEDED

- pencils
- objects for counting
- a new set of objects for counting—ten strips of colored paper (a different color from the set made in LIFEPAC 101, page 2), one inch by five inches in size, pasted or glued to cardboard—write the number 10 on each one of these new objects
- chart of numbers

Concept:

number order to 50

Teacher Goal:

To teach the children to write the numbers in order from 1 to 50.

Teaching Page 20:

Place ten of the original strips in front of the students and then give them one of the new strips. Explain to the students that these sets have the same value. Continue to do this until the students understand that *two* of the new objects represents *20* and so on. Place one of the *tens'* objects and three of the *ones'* objects in front of the students and ask them to say the number. (thirteen)

Continue this exercise until the children understand the new value of the *tens'* objects and the *ones'* objects for counting. Say a number aloud to the students and have them select the correct objects to form the number. This exercise serves two purposes. It reinforces the concept of *tens* and *ones* and it simplifies counting with objects. Use the chart of numbers to count to *99* aloud.

Turn to page 20 and ask the children if the officers are standing in number order. Read the instructions aloud. The students may use the chart of numbers to complete the problems if necessary.

ADD/SUBTRACT TO 10, SHAPES | Unit 2

Write the numbers in order.

7   2   6   3   4   5
_2_ , _3_ , _4_ , _5_ , _6_ , _7_

15   17   16   14   19   18
_14_ , _15_ , _16_ , _17_ , _18_ , _19_

23   25   21   26   22   24
_21_ , _22_ , _23_ , _24_ , _25_ , _26_

36   38   40   41   39   37
_36_ , _37_ , _38_ , _39_ , _40_ , _41_

45   43   46   42   44   47
_42_ , _43_ , _44_ , _45_ , _46_ , _47_

20 | Section 4

PAGE 21: NUMBER ORDER

MATERIALS NEEDED

- pencils
- chart of numbers
- objects for counting (ones and tens)

Concept:

number order to 99

Teacher Goal:

To teach the children to write numbers in order from 50 to 99.

Teaching Page 21:

Have the children use the chart of numbers to count to *99*. Have them count to *99* aloud without using the chart. Point to any two numbers on the chart and ask which number is *bigger* and which number is *smaller*. Students who have difficulty should use the objects for counting to illustrate the numbers with sets. Make sets of *ten*, *twenty*, *thirty*, and so on. Have the children point to those numbers on the chart of numbers. Talk about *bigger* and *smaller*.

Turn to page 21 and read the instructions. Tell the children that number order is arranging numbers from the *smallest* to the *biggest*. The children may use objects for counting or the chart of numbers to complete this page.

Write the numbers in order.

52	53	50	48	49	51
48	49	50	51	52	53

62	65	61	60	64	63
60	61	62	63	64	65

69	73	71	74	72	70
69	70	71	72	73	74

83	80	84	85	82	81
80	81	82	83	84	85

90	87	91	86	89	88
86	87	88	89	90	91

Section 4 | **21**

PAGES 22 & 23: NUMBER ORDER

MATERIALS NEEDED

• pencils

Concept:

number order to 99

Teacher Goal:

To teach the children to identify numbers that are greater than or less than another number.

Teaching Pages 22 and 23:

Turn to page 22 and read the instructions to the children. Explain that they should first write the numbers in number order on the lines. Then they should circle the numbers that are greater than *10*, *24*, *45*, *30*, and *72*. Students should be encouraged to look at the number in the tens' position to help recognize number order and compare number size.

Turn to page 23 and read the instructions to the children. Explain that they should first write the numbers in number order on the lines. Then they should circle the numbers that are less than *10*, *36*, *52*, *75*, and *99*.

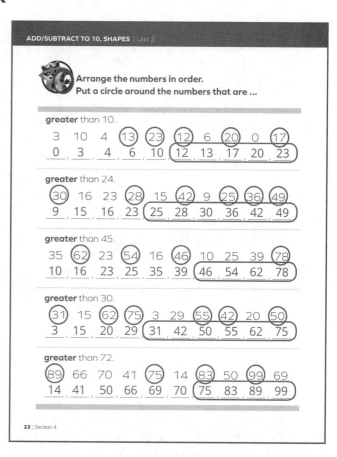

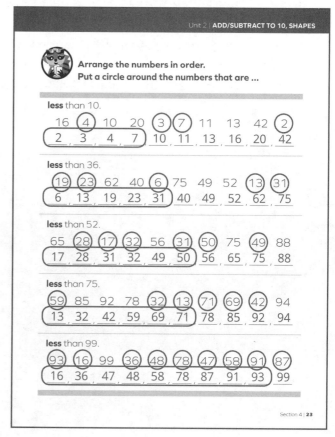

PAGE 24: MEASUREMENTS

MATERIALS NEEDED

- pencils
- paper
- ruler used in LIFEPAC 101, page 25
- scale for weighing
- book
- quart of milk or similar liquid
- bag of flour
- jar containing small objects
- cups
- medium-size box
- blocks

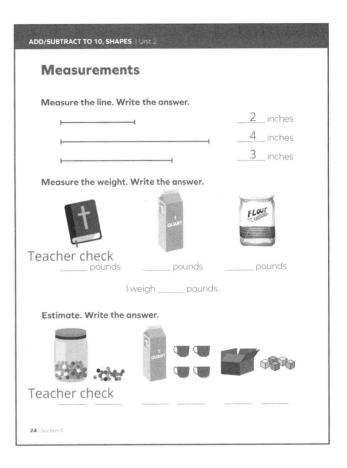

ADD/SUBTRACT TO 10, SHAPES | Unit 2

Measurements

Measure the line. Write the answer.

_____ 2 inches
_____ 4 inches
_____ 3 inches

Measure the weight. Write the answer.

Teacher check
_____ pounds   _____ pounds   _____ pounds

I weigh _____ pounds.

Estimate. Write the answer.

Teacher check
_____ _____

24 | Section 4

Concept:

measurements—length, weight, "how many"

Teacher Goals:

To teach the children to measure in inches and pounds and to estimate "how many."

Teaching Page 24:

Write the word *measure* on the board and ask the children if they remember what it means to measure. Remind them that there are many things that can be measured. Place the materials that have been gathered in front of the children and identify each one. Ask them which objects are used to take measurements. (ruler, scale) Discuss the *inches* on the ruler and introduce *pounds* on the scale. Using the ruler, paper, and pencil, have the children draw lines of three inches, five inches, and one inch.

Turn to page 24 and read the first set of directions aloud. Tell the children to use their rulers to measure the lines in inches and write the answer. Read the next set of directions. Explain that this time the children will be measuring how heavy. Have them measure the weight of the book, the weight of the quart of liquid, and the weight of the bag of flour. Be sure that they use and understand the word *pounds*. Allow the students to measure and record their own weight. Ask them to compare their weight to the other objects they have measured.

Introduce the word *estimate*. Explain to the children that they cannot always measure an object and so they must learn to estimate its size. Place the jar that contains the small objects in front of the children. Ask them to estimate (to tell how many) objects they think are in the jar and write the number on the first line. Have them empty the jar and count the objects. Tell them to write this number on the second line. Talk about which number is bigger or smaller, how close the students were in their estimation, and why they gave the answer that they did. Follow the same procedure in estimating the number of cups in the quart of liquid, and the number of blocks in a medium-size box.

PAGE 25: SHAPES

MATERIALS NEEDED

- pencils
- paper
- scissors
- square box (or block)
- ball (or round balloon)
- glass
- rectangular box
- ice cream cone (funnel or cone made from paper)

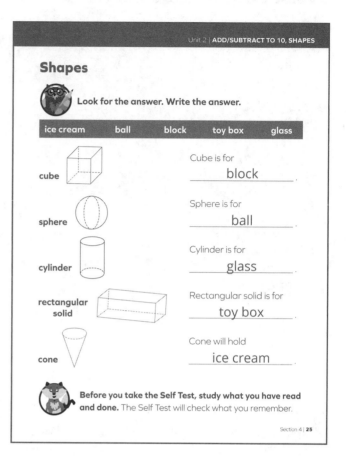

Concept:

solid shapes

Teacher Goals:

To teach the children to recognize solid shapes and to learn about cube, sphere, cylinder, rectangular solid, and cone.

Teaching Page 25:

Let the children use the paper and scissors to cut out familiar shapes of circles, squares, rectangles, and triangles. Be sure that they can identify each one. Point out that these are all flat shapes. They cannot put anything into these shapes.

Introduce the expression *solid shapes*.

Turn to page 25. Have the children point to the illustration and then the word next to it. Say each word aloud with them. Tell the children that these are examples of solid shapes. (In math, we use the term *solid*, although some objects are not really solid but have an inside and out-side.) Place the materials for this lesson in front of the children and have them identify each object with one of the new words on page 25. Talk about inside and outside and illustrate by putting things inside solid shapes (air in balloon, water in glass). Read the directions and the words at the top of the page 25. Read the sentences aloud with the children and help them select the correct answer.

The students should prepare for the Self Test. Ask the students to look over and read the Self Test but they should not write the answers to any questions. After looking over the Self Test the students should go to the beginning of the unit and reread the text and review the answers to the activities up to the Self Test.

The students are to complete the Self Test the next school day. This should be done under regular test conditions without allowing the students to look back. A good idea is to clip the pages together before the test.

PAGE 26: SELF TEST 4

MATERIALS NEEDED

• pencils

Concepts:

number order to 99, measurements, shapes

Teacher Goal:

To teach the children to learn to check their progress periodically.

Teaching Page 26:

Turn to page 26. Read the directions to the children. Have the children repeat them after you while running their fingers under the sentence being read. Be sure the children understand what they are to do. If necessary, students may use the chart of numbers. Let the children complete the page. You may repeat the directions but give no other help.

Do not have the children check their own work. Check it as soon as you can, and go over it with each child. Show him where he did well and where he needs extra help.

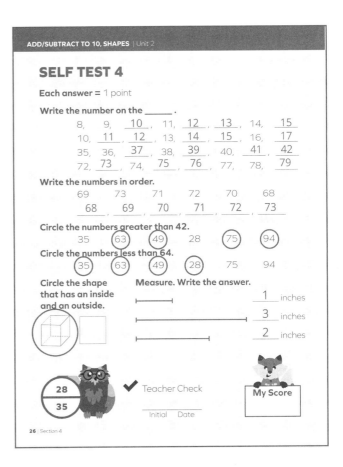

5. NUMBERS AS WORDS

PAGE 27: NUMBER WORDS

MATERIALS NEEDED

• pencils

Concept:

number words *zero* to *ten*

Teacher Goals:

To teach the children to read and write the number words *zero* through *ten* and to match number words and number symbols.

Teaching Page 27:

Turn to page 27. Ask the students to read the number symbols and the number words aloud. Read the directions with them and have them complete the first exercise. Read the next set of directions and allow them to complete the page independently.

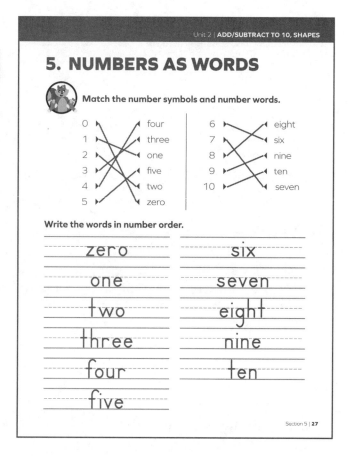

PAGE 28: NUMBER WORDS

MATERIALS NEEDED

- pencils
- paper

Concept:

number facts in words

Teacher Goal:

To teach the children to write the number facts in number words and number symbols, vertically and horizontally.

Teaching Page 28:

Dictate the following problems to the students. Have the students write the problems on a piece of paper in number symbols—first vertically (up and down) and second horizontally (across). Then have them write each problem in number words using operations symbols (+, −, =).

1. Three + two = five.
2. Seven − one = six.
3. Four + three = seven.

one two three four five six seven eight nine ten

Write the number facts.

five + three = eight	seven − four = three
5 + 3 = 8	7 − 4 = 3
eight + two = ten	nine − three = six
8 + 2 = 10	9 − 3 = 6

Write the number words.

7
+ 1
8   Seven + one = eight.

6
− 2
4   Six − two = four.

3
+ 3
6   Three + three = six.

9
− 5
4   Nine − five = four.

Write the number word.

Two + seven = nine.	Six − five = one.
Three + five = eight.	Seven − two = five.

28 | Section 5

Turn to page 28 and ask the children to read aloud the words at the top of the page. Ask them to read the first number fact aloud and then write the fact on the lines in number symbols. Have the children complete the first exercise. Read the next set of directions aloud. Explain to the children that this time they should write the number word on the line. Allow them to complete this exercise. Read the last set of directions. Tell the children that you are going to dictate four problems to them and that they should write the problem in the box using number words and operation symbols.

1. Two + seven = nine.
2. Six − five = one.
3. Three + five = eight.
4. Seven − two = five.

PAGE 29: ORDINAL NUMBERS

MATERIALS NEEDED

- pencils
- objects for counting
- small box

Concept:

number order

Teacher Goal:

To teach the children to learn number order by using ordinals numbers first through tenth.

Teaching Page 29:

Place *ten* objects in a row and have the students count from *one* to *ten*. Explain to them that we have another way to express number order. Have them point to the objects and say aloud with them, "This is the first object in the row, this is the second object in the row," and so on until they reach the *tenth object*. Place the box in front of the student and tell him to select the *first object* and place it in the box, then the *second object*. Continue doing this until the *tenth object* is in the box.

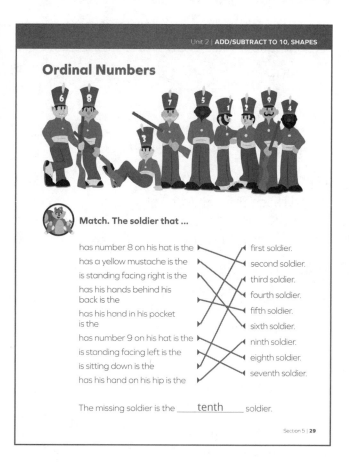

Turn to page 29 and have the children point to the soldiers at the top of the page. Talk to the students about the soldiers—how they are standing, the numbers on their hats, their beards and mustaches, etc.

Starting from the left and moving to the right, have the children identify the soldiers as the first soldier, the second soldier, the third soldier, and so on until they identify the ninth soldier. (the one with the 4 on his hat) Read the directions and the first sentence aloud. "The soldier that has number 8 on his hat is the ...." Have the children locate the soldier with the number 8 on his hat in the picture. Then have the students repeat the order first soldier, second soldier. When they identify the soldier with the number 8 on his hat as the second soldier have them draw a matching line. This page should be completed orally with the students.

PAGE 30: PATTERNS AND SEQUENCING

MATERIALS NEEDED

- pencils
- crayons
- big and little objects

Concept:

patterns and sequencing

Teacher Goals:

To teach the children to learn to identify patterns using numbers and illustrations and to tell what comes next.

Teaching Page 30:

Place a group of objects in front of the students and have them separate them into a group of *big* objects and a group of *little* objects. Arrange several of the objects in a line so that they form a pattern of *big*, *little*, *big*, *little*. Ask the students to identify the pattern and tell what the next object would be. Make another arrangement of two *little*, one *big*, two *little*, one *big*. Have the students identify the pattern and tell them what should be next. Place the crayons in a pattern of *up*, *down*, *up*, *down*, and have the students explain that pattern. Allow the students to make several patterns of their own and have other students identify the patterns and tell what comes next.

Turn to page 30 and read the directions at the top of the page. Help the children complete the first exercise and then allow them to finish the page independently. When all exercises are complete have the students describe each one of the patterns aloud and explain their answers.

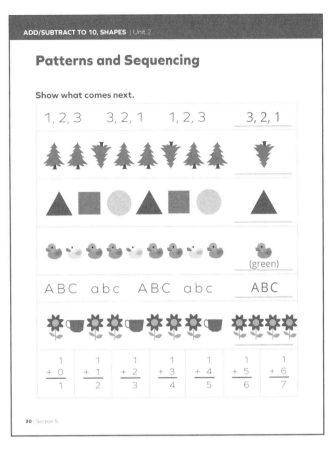

PAGE 31: COLORS AND DETAIL

MATERIALS NEEDED

- pencils
- crayons

Concept:

colors and details

Teacher Goals:

To teach the children to review the names of the colors and to identify details in a picture.

Teaching Page 31:

Review colors with the students using the crayons. Students should be able to identify red, yellow, green, blue, brown, purple, orange, black, white, and pink.

Turn to page 31 and read aloud the rhyme at the top of the page with the children. Tell them to count the pigs and the ears of corn. Ask if they think there are enough ears for all the pigs.

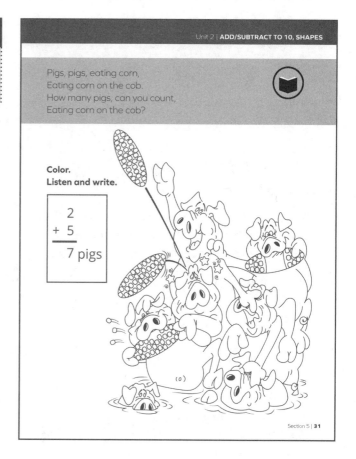

Tell the children to listen and write a fact in the box. There are two pigs eating corn and five that are not. How many pigs are there altogether? (2 + 5 = 7)

Identify the stars in the picture and ask why they are there. What do the small semi-circles at the bottom of the page represent? Have the students color the picture. Encourage them to use a different color for each pig. Tell them use their colors to show as much detail in the picture as possible.

PAGE 32: NUMBER FACTS

MATERIALS NEEDED

- pencils
- fact cards for addition and subtraction
- chart of numbers

Concepts:

counting, number facts to 10

Teacher Goals:

To teach the children to count to 99 and to review number facts to 10.

Teaching Page 32:

Review addition and subtraction fact cards with the students.

Turn to page 32 and read the directions at the top of the page. This is review work and the students should be able to complete the page independently.

Be sure they understand that in the first exercise, each line is a new number grouping. Children should use the manipulatives only as necessary.

ADD/SUBTRACT TO 10, SHAPES | Unit 2

Write the number in the □.

10	11	12	**13**	14	15	16	17	18	**19**
40	41	**42**	43	44	45	**46**	47	48	49
50	51	52	53	54	**55**	56	57	58	59
60	61	62	**63**	64	65	66	67	**68**	69
80	**81**	82	83	**84**	85	86	87	88	89
90	91	92	**93**	94	95	96	**97**	98	99

$4 + 6 = \underline{10}$ $3 - 1 = \underline{2}$ $10 + 3 = \underline{13}$

$3 + 4 = \underline{7}$ $4 - 2 = \underline{2}$ $8 + 4 = \underline{12}$

$5 + 5 = \underline{10}$ $5 - 3 = \underline{2}$ $6 - 5 = \underline{1}$

$4 + 0 = \underline{4}$ $7 - 2 = \underline{5}$ $9 - 4 = \underline{5}$

| $\begin{array}{r} 10 \\ +\ 1 \\ \hline 11 \end{array}$ | $\begin{array}{r} 10 \\ +\ 4 \\ \hline 14 \end{array}$ | $\begin{array}{r} 10 \\ +\ 5 \\ \hline 15 \end{array}$ | $\begin{array}{r} 10 \\ +\ 3 \\ \hline 13 \end{array}$ | $\begin{array}{r} 10 \\ -\ 2 \\ \hline 8 \end{array}$ | $\begin{array}{r} 10 \\ -\ 8 \\ \hline 2 \end{array}$ | $\begin{array}{r} 10 \\ -\ 6 \\ \hline 4 \end{array}$ | $\begin{array}{r} 10 \\ -\ 7 \\ \hline 3 \end{array}$ |

32 | Section 5

PAGE 33: NUMBER FACTS

MATERIALS NEEDED

- pencils
- fact cards for addition and subtraction
- objects for counting (ones)

Concept:

addition and subtraction facts

Teacher Goal:

To teach the children to learn that different number facts have the same answer.

Teaching Page 33:

Review the fact cards to ten for addition and subtraction. Tell the children you are going to play a game. Tell them that you are going to give them a group of objects and that they must tell you how many facts they can find in that group of objects. Begin with *four* objects. The students should be able to identify the facts (in any order) of 0 + 4, 1 + 3, 2 + 2, 3 + 1, 4 + 0. Give the students *six* objects and have them find at least *four* facts. Give the students *eight* objects and have them identify at least *three* facts. Continue until the students understand that a number may represent several addition facts.

Now reverse the process using subtraction. Give the students *three* objects. Ask them to illustrate a subtraction fact using objects for counting that results in an answer of *three* (9 − 6, 8 − 5, 7 − 4). Continue this exercise until the students understand that many subtraction facts may result in the answer of *3, 4, 5,* and so on.

Turn to page 33. Read the directions at the top of the page. Explain to the students that the answer is given to them. They must write two number facts for each answer. Tell them they may write any fact that they want providing the answer is correct. Have the students complete the page independently. They may use objects for counting and fact cards if necessary.

The students should prepare for the Self Test. Ask the students to look over and read the Self Test but they should not write the answers to any questions. After looking over the Self Test the students should go to the beginning of the unit and reread the text and review the answers to the activities up to the Self Test.

The students are to complete the Self Test the next school day. This should be done under regular test conditions without allowing the students to look back. A good idea is to clip the pages together before the test.

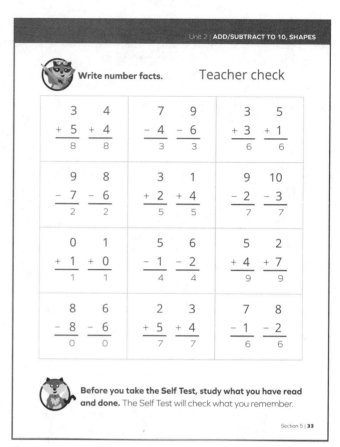

PAGE 34: SELF TEST 5

MATERIALS NEEDED

• pencils
• crayons

Concepts:

number words, facts, sequence, ordinal numbers, story problems

Teacher Goal:

To teach the children to learn to check their progress periodically.

Teaching Page 34:

Turn to page 34. Read the directions to the children. Have the children repeat them after you while running their fingers under the sentence being read. Be sure the children understand what they are to do. You may repeat the directions but give no other help. Dictate the following:

1. Listen and write:

    Three + four = seven.

    Eight – two = six.

    (Students should write number facts in words.)

2. Listen and write:

    Color the fourth cat blue, the third cat green, the sixth cat red.

Write the number fact for the story:

Mary and Jack went to the pet store and saw three white rabbits and two brown rabbits. How many rabbits were there altogether?

(Students must write the answer as a number fact in symbols.)

Do not have the children check their own work. Check it as soon as you can, and go over it with each child. Show him where he did well and where he needs extra help.

The students should prepare for the LIFEPAC Test. The students should go to the beginning of the unit and reread the text and review the answers to the activities for the entire unit. Ask the students questions to check their understanding of the unit.

The students are to complete the LIFEPAC Test the next school day. This should be done under regular test conditions without allowing the students to look at the unit.

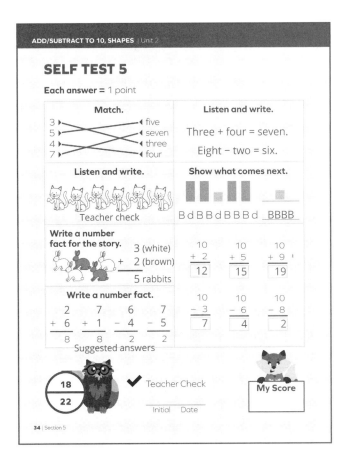

LIFEPAC TEST 102

MATERIALS NEEDED

- pencils
- crayons
- pennies

Concepts:

add and subtract to 10 vertically and horizontally, place value for tens and ones, add to 10, solid and flat shapes, number words, sequence, number order, solve oral story problems

Teacher Goal:

To teach the children to learn to check their own progress periodically.

Teaching the LIFEPAC Test:

Administer the test in at least two sessions.

Read all of the directions on each page as the children prepare to do it. Be sure that they understand what they are being asked to do.

LIFEPAC Test page 4

Listen and write.

(Color and selection should be correct.)
    Circle ...

        the *fifth* apple yellow,

        the *second* apple black,

        the *third* apple red.

Write the number fact for the story:

    Mary and Jack went to the store. They bought six sugar cookies and four chocolate cookies. How many cookies did they buy altogether?

(Students must write the answer as a *number fact* in number words.)

Give no help except with directions.

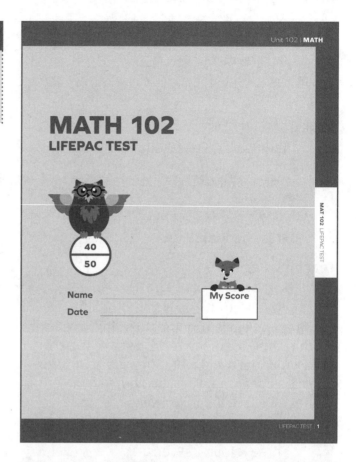

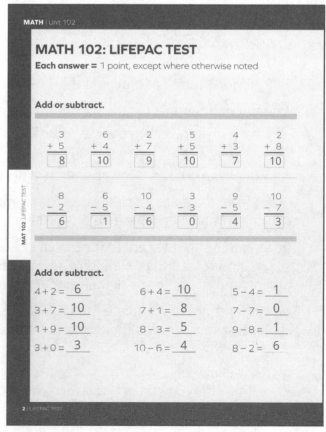

Go over each page with the child as soon as possible after you check it so that he can see where he did well and where he needs more work.

Evaluate the tests and review areas where the children have done poorly. Review the pages and activities that stress the concepts tested.

If necessary, when the children have reviewed sufficiently, administer the Alternate LIFEPAC test. Follow the same procedures as used for the LIFEPAC Test.

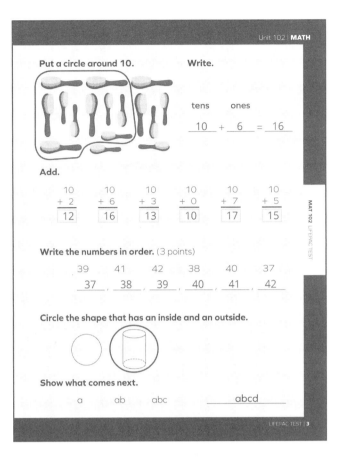

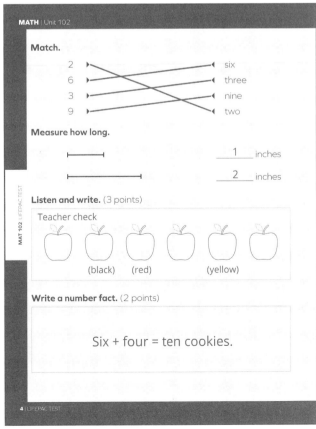

ALTERNATE LIFEPAC TEST 102

MATERIALS NEEDED

- pencils
- crayons
- pennies

Concepts:

add and subtract to 10 vertically and horizontally, place value for tens and ones, add to 10, solid and flat shapes, number words, sequence, number order, solve oral story problems

Teacher Goal:

To teach the children to learn to check their own progress periodically.

Teaching the Alternate LIFEPAC Test:

Administer the test in at least two sessions.

Read all of the directions on each page as the children prepare to do it. Be sure that they understand what they are being asked to do.

Alternate LIFEPAC Test page 4

Listen and write.

(Color and selection should be correct.)

Circle ...

the *first* car blue,

the *sixth* car green,

the *fourth* car orange.

Write the number fact for the story:

Jerry had five pennies in one pocket and four pennies in another pocket. How many pennies did he have altogether?

(Students must write the answer as a *number fact* in number words.)

Give no help except with directions.

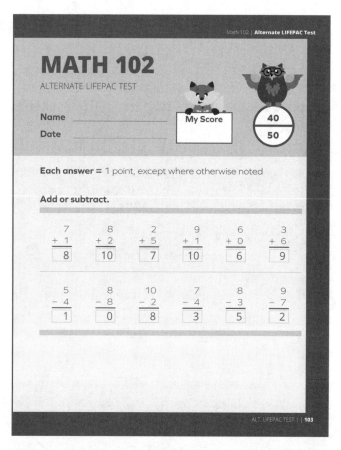

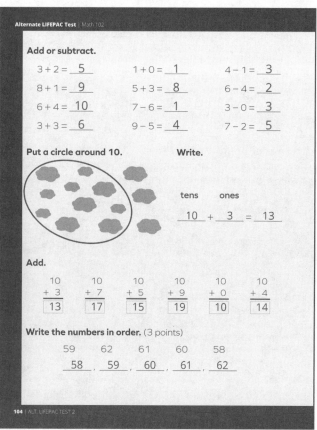

Go over each page with the child as soon as possible after you check it so that he can see where he did well and where he needs more work.

Evaluate the tests and review areas where the children have done poorly. Review the pages and activities that stress the concepts tested.

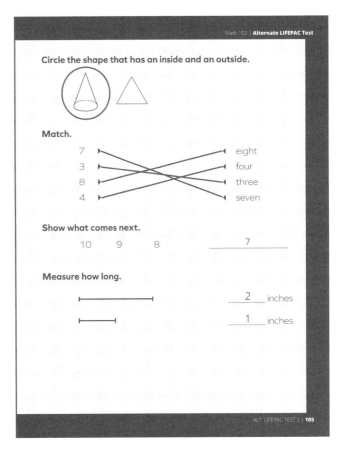

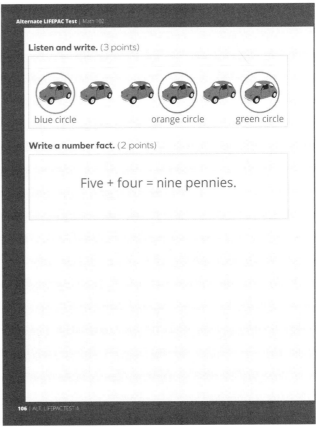

MATH 102

ALTERNATE LIFEPAC TEST

Name _____

Date _____

My Score

40
___
50

Each answer = 1 point, except where otherwise noted

Add or subtract.

7	8	2	9	6	3
+ 1	+ 2	+ 5	+ 1	+ 0	+ 6
☐	☐	☐	☐	☐	☐

5	8	10	7	8	9
− 4	− 8	− 2	− 4	− 3	− 7
☐	☐	☐	☐	☐	☐

Add or subtract.

3 + 2 = _____   1 + 0 = _____   4 − 1 = _____

8 + 1 = _____   5 + 3 = _____   6 − 4 = _____

6 + 4 = _____   7 − 6 = _____   3 − 0 = _____

3 + 3 = _____   9 − 5 = _____   7 − 2 = _____

Put a circle around 10.

Write.

tens   ones

_____ + _____ = _____

Add.

| 10 | 10 | 10 | 10 | 10 | 10 |
| + 3 | + 7 | + 5 | + 9 | + 0 | + 4 |

Write the numbers in order. (3 points)

59   62   61   60   58

_____ , _____ , _____ , _____ , _____

Circle the shape that has an inside and an outside.

Match.

7 ▶   ◀ eight

3 ▶   ◀ four

8 ▶   ◀ three

4 ▶   ◀ seven

Show what comes next.

10   9   8   _____

Measure how long.

├────────────┤   _____ inches

├──────┤   _____ inches

Listen and write. (3 points)

Write a number fact. (2 points)

MATH 103

Unit 3: Fractions, Time, and Symbols

FRACTIONS, TIME, AND SYMBOLS
MATH 103

Alpha Omega
PUBLICATIONS

804 N. 2nd Ave. E.
Rock Rapids, IA 51246-1759

© MCMXCVII by Alpha Omega Publications, Inc.
All rights reserved.
LIFEPAC is a registered trademark of Alpha
Omega Publications, Inc.

Author:
Carol Bauler, B.A.

Editor:
Alan Christopherson, M.S.

Media Credits:
Page 1: © wenchiawang, iStock, Thinkstock.
19: © Gurzzza, iStock, Thinkstock; © Elvetica,
iStock, Thinkstock; © Askold Romanov, Hemera,
Thinkstock, PrettyVectors, iStock, Thinkstock;
32: © Allevinatis, iStock, Thinkstock; © macoo,
iStock, Thinkstock; **34, LP3:** © kazberry, iStock,
Thinkstock.

i

PAGE 1: FRACTIONS, TIME, AND SYMBOLS

MATERIALS NEEDED

• pencils

Concepts:

purpose of LIFEPAC, objectives

Teacher Goals:

To teach the children to know what is expected of the student in the LIFEPAC and to write first and last names correctly.

Teaching Page 1:

Turn to page 1. Point to the title and read it aloud. Allow time for the children to look through the LIFEPAC. Write the word *Objectives* on the board and have the children find the word on the page. Explain that the objectives tell the things the students will be expected to do in the LIFEPAC. Read each one and have the children repeat them as they run their fingers under the sentence from left to right. Talk about the objectives so that the children will understand what they will be doing. Have each child write his name on the line.

Unit 3 | FRACTIONS, TIME, AND SYMBOLS

FRACTIONS, TIME, AND SYMBOLS

Objectives

Read these objectives. They will tell what you will be able to do when you have finished this LIFEPAC®.

1. I can read addition and subtraction number sentences.
2. I can recognize the signs for = and ≠.
3. I can find fractions in shapes and objects.
4. I can follow oral directions.
5. I can read the calendar and tell time to the hour.
6. I can use my math skills to solve story problems.

My name is

Introduction | **1**

1. ADD AND SUBTRACT TO 10

PAGE 2: FACTS TO 10

MATERIALS NEEDED

- pencils
- objects for counting
- addition fact cards

Concept:

addition facts to 10

Teacher Goal:

To teach the children to add number facts to 10 vertically.

Teaching Page 2:

Turn to page 2. Explain to the children that the facts on this page are just like their fact cards. Tell the children that this exercise will let them know how many of their addition facts they have learned.

Set a reasonable time limit for the page to be completed. Correct the page with the students. Select fact cards for those problems that were incorrect and review. Use objects for counting if necessary.

FRACTIONS, TIME, AND SYMBOLS | Unit 3

1. ADD AND SUBTRACT TO 10

Write the number in the ☐.

1	2	2	3	4	5	5	1	8
+8	+4	+1	+5	+0	+0	+5	+6	+0
9	6	3	8	4	5	10	7	8
5	4	3	3	6	5	8	9	2
+2	+2	+0	+6	+1	+4	+1	+1	+2
7	6	3	9	7	9	9	10	4
1	3	5	4	7	9	4	1	5
+9	+7	+1	+3	+2	+0	+1	+3	+3
10	10	6	7	9	9	5	4	8
1	1	4	8	2	3	1	7	3
+4	+0	+5	+2	+5	+1	+1	+1	+2
5	1	9	10	7	4	2	8	5
2	2	1	2	4	1	3	1	6
+6	+0	+2	+2	+6	+5	+4	+7	+3
8	2	3	4	10	6	7	8	9
7	6	7	4	2	6	3	2	6
+0	+4	+3	+4	+8	+0	+3	+3	+2
7	10	10	8	10	6	6	5	8

2 | Section 1

PAGE 3: FACTS TO 10

MATERIALS NEEDED

- pencils
- objects for counting
- subtraction fact cards

Concept:

subtraction facts to 10

Teacher Goal:

To teach the children to subtract number facts to 10 vertically.

Teaching Page 3:

Turn to page 3. Explain to the children that the facts on this page are just like their fact cards. Tell the children that this exercise will let them know how many of their subtraction facts they have learned.

Set a reasonable time limit for the page to be completed. Correct the page with the students. Select fact cards for those problems that were incorrect and review. Use objects for counting if necessary.

Unit 3 | **FRACTIONS, TIME, AND SYMBOLS**

Write the number in the ☐ .

3 − 1 = 2	6 − 4 = 2	3 − 2 = 1	1 − 0 = 1	9 − 6 = 3	5 − 2 = 3	6 − 0 = 6	7 − 1 = 6	8 − 3 = 5
4 − 2 = 2	8 − 6 = 2	10 − 8 = 2	4 − 1 = 3	9 − 5 = 4	7 − 3 = 4	8 − 1 = 7	9 − 0 = 9	7 − 4 = 3
5 − 3 = 2	7 − 5 = 2	8 − 7 = 1	6 − 5 = 1	5 − 1 = 4	7 − 2 = 5	8 − 5 = 3	10 − 4 = 6	4 − 0 = 4
4 − 3 = 1	7 − 6 = 1	10 − 9 = 1	10 − 1 = 9	9 − 4 = 5	3 − 0 = 3	5 − 4 = 1	2 − 1 = 1	10 − 5 = 5
9 − 1 = 8	9 − 3 = 6	9 − 2 = 7	7 − 0 = 7	8 − 2 = 6	5 − 0 = 5	9 − 7 = 2	6 − 3 = 3	6 − 4 = 2
2 − 0 = 2	9 − 8 = 1	10 − 7 = 3	6 − 2 = 4	8 − 4 = 4	10 − 2 = 8	10 − 6 = 4	8 − 0 = 8	10 − 3 = 7

Section 1 | **3**

PAGE 4: FACTS TO 10

MATERIALS NEEDED

- pencils
- objects for counting
- addition fact cards
- number symbol cards

Concept:

addition facts to 10

Teacher Goal:

To teach the children to add number facts to 10 horizontally.

Teaching Page 4:

Turn to page 4 and have the children read the directions at the top of the page. Ask them to tell you what is different about these facts compared to page 2 (horizontal instead of vertical). Point to several of the problems on the page and ask the students to read the problems aloud. Remind them that the answers to the facts are the same—horizontally or vertically.

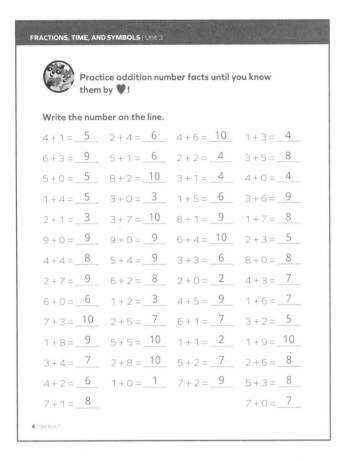

FRACTIONS, TIME, AND SYMBOLS | Unit 3

Practice addition number facts until you know them by ♥!

Write the number on the line.

4 + 1 = 5	2 + 4 = 6	4 + 6 = 10	1 + 3 = 4
6 + 3 = 9	5 + 1 = 6	2 + 2 = 4	3 + 5 = 8
5 + 0 = 5	8 + 2 = 10	3 + 1 = 4	4 + 0 = 4
1 + 4 = 5	3 + 0 = 3	1 + 5 = 6	3 + 6 = 9
2 + 1 = 3	3 + 7 = 10	8 + 1 = 9	1 + 7 = 8
9 + 0 = 9	9 + 0 = 9	6 + 4 = 10	2 + 3 = 5
4 + 4 = 8	5 + 4 = 9	3 + 3 = 6	8 + 0 = 8
2 + 7 = 9	6 + 2 = 8	2 + 0 = 2	4 + 3 = 7
6 + 0 = 6	1 + 2 = 3	4 + 5 = 9	1 + 6 = 7
7 + 3 = 10	2 + 5 = 7	6 + 1 = 7	3 + 2 = 5
1 + 8 = 9	5 + 5 = 10	1 + 1 = 2	1 + 9 = 10
3 + 4 = 7	2 + 8 = 10	5 + 2 = 7	2 + 6 = 8
4 + 2 = 6	1 + 0 = 1	7 + 2 = 9	5 + 3 = 8
7 + 1 = 8		7 + 0 = 7	

4 | Section 1

Set a reasonable time limit for the page to be completed. Correct the page with the students and compare the answers to their answers on page 2. Select fact cards for those problems that were incorrect and review with the students. Select a group of addition fact cards and number symbol cards and play a game of concentration.

PAGE 5: FACTS TO 10

MATERIALS NEEDED

- pencils
- objects for counting
- subtraction fact cards
- number symbol cards

Concept:

subtraction facts to 10

Teacher Goal:

To teach the children to subtract number facts to 10 horizontally.

Teaching Page 5:

Turn to page 5 and have the children read the directions at the top of the page. Ask them to tell you what is different about these facts compared to page 3 (horizontal instead of vertical). Point to several of the problems on the page and ask the students to read the problems aloud. Remind them that the answers to the facts are the same— horizontally or vertically.

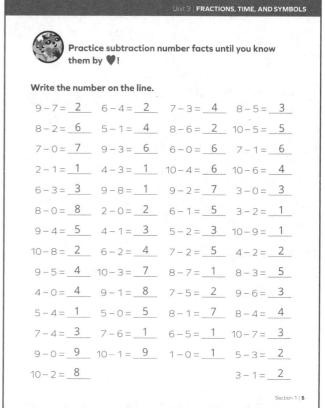

Unit 3 | **FRACTIONS, TIME, AND SYMBOLS**

Practice subtraction number facts until you know them by ♥!

Write the number on the line.

9 − 7 = 2	6 − 4 = 2	7 − 3 = 4	8 − 5 = 3
8 − 2 = 6	5 − 1 = 4	8 − 6 = 2	10 − 5 = 5
7 − 0 = 7	9 − 3 = 6	6 − 0 = 6	7 − 1 = 6
2 − 1 = 1	4 − 3 = 1	10 − 4 = 6	10 − 6 = 4
6 − 3 = 3	9 − 8 = 1	9 − 2 = 7	3 − 0 = 3
8 − 0 = 8	2 − 0 = 2	6 − 1 = 5	3 − 2 = 1
9 − 4 = 5	4 − 1 = 3	5 − 2 = 3	10 − 9 = 1
10 − 8 = 2	6 − 2 = 4	7 − 2 = 5	4 − 2 = 2
9 − 5 = 4	10 − 3 = 7	8 − 7 = 1	8 − 3 = 5
4 − 0 = 4	9 − 1 = 8	7 − 5 = 2	9 − 6 = 3
5 − 4 = 1	5 − 0 = 5	8 − 1 = 7	8 − 4 = 4
7 − 4 = 3	7 − 6 = 1	6 − 5 = 1	10 − 7 = 3
9 − 0 = 9	10 − 1 = 9	1 − 0 = 1	5 − 3 = 2
10 − 2 = 8			3 − 1 = 2

Section 1 | **5**

Set a reasonable time limit for the page to be completed. Correct the page with the students and compare the answers to their answers on page 3. Select fact cards for those problems that were incorrect and review with the student. Select a group of subtraction fact cards and number symbol cards and play a game of concentration.

PAGE 6: FACTS TO 10

MATERIALS NEEDED

• pencils
• objects for counting

Concept:

addition and subtraction facts to 10

Teacher Goal:

To teach the children to add and subtract number facts to 10 horizontally.

Teaching Page 6:

Turn to page 6. Explain to the children that they will be "teacher" today and will grade another student's work. Read the directions at the top of the page, and allow the students to complete the page independently. Correct the page with the students and congratulate them if they have been good "teachers."

The students should prepare for the Self Test. Ask the students to look over and read the Self Test but they should not write the answers to any questions. After looking over the Self Test the students should go to the beginning of the unit and reread the text and review the answers to the activities up to the Self Test.

The students are to complete the Self Test the next school day. This should be done under regular test conditions without allowing the students to look back. A good idea is to clip the pages together before the test.

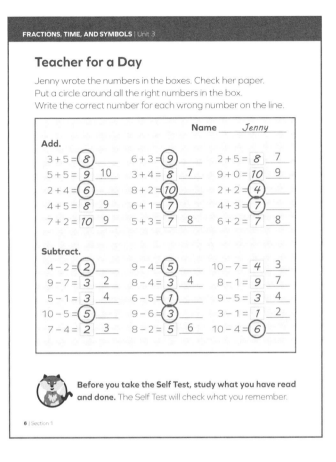

PAGE 7: SELF TEST 1

MATERIALS NEEDED

- pencils
- addition and subtraction fact cards

Concept:

addition and subtraction facts

Teacher Goal:

To teach the children to learn to check their progress periodically.

Teaching Page 7:

Turn to page 7. Read the directions to the children. Have the children repeat them after you while running their fingers under the sentence being read. Be sure the children understand what they are to do. You may repeat the directions but give no other help.

Do not have the children check their own work. Check it as soon as you can, and go over it with each child. Show him where he did well and where he needs extra help.

Continue to review the fact cards regularly. Present the cards in order and with objects for counting to students who are having difficulty. Scramble the cards for students who are more advanced.

Unit 3 | FRACTIONS, TIME, AND SYMBOLS

SELF TEST 1

Each answer = 1 point

Write the number in the ☐ or on the _____ .

1 + 2 **3**	3 + 6 **9**	5 + 0 **5**	2 + 5 **7**	3 + 4 **7**	5 + 3 **8**
4 + 5 **9**	2 + 8 **10**	3 + 3 **6**	2 + 2 **4**	5 + 5 **10**	7 + 2 **9**
7 − 4 **3**	6 − 0 **6**	9 − 3 **6**	4 − 2 **2**	8 − 6 **2**	9 − 4 **5**
8 − 3 **5**	10 − 4 **6**	3 − 2 **1**	9 − 6 **3**	8 − 2 **6**	6 − 2 **4**

6 + 3 = __9__   6 − 2 = __4__
1 + 9 = __10__   10 − 3 = __7__
3 + 5 = __8__   9 − 6 = __3__
2 + 8 = __10__   7 − 3 = __4__
5 + 4 = __9__   8 − 5 = __3__

27
—
34

✔ Teacher Check

Initial   Date

My Score

Section 1 | **7**

2. READING A NUMBER SENTENCE

PAGES 8 & 9:
NUMBER SENTENCES

MATERIALS NEEDED

• pencils

Concept:

reading number sentences

Teacher Goal:

To teach the children to read number sentences in addition and subtraction.

Teaching Pages 8 and 9:

Turn to page 8. Tell the children that a sentence is a statement that tells them something. Explain to them that they can make a statement using numbers and/or words. Have the children point to the sets of pans and then have them point to the number fact. Ask them what the operation sign (+) stands for (plus) and what the sign (=) stands for (equals). Have the children point to each word as they read the sentence, "Three plus four equals seven." Read the directions and have the children complete the page.

Turn to page 9. Remind the children that a sentence is a statement that tells them something. It may be in numbers or in words. Have the students point to the sets of hearts and then have them point to the number fact. Ask them what the operation sign (–) stands for (minus) and what the sign (=) stands for (equals). Have the children point to each word as they read the sentence, "Five minus two equals three." Read the directions and have the children complete the page.

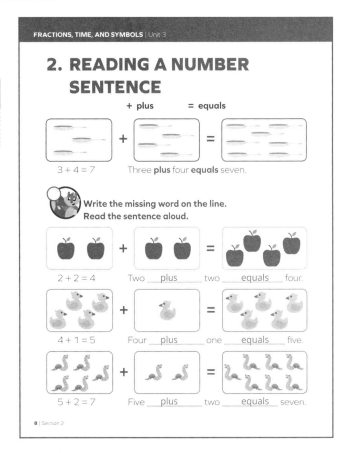

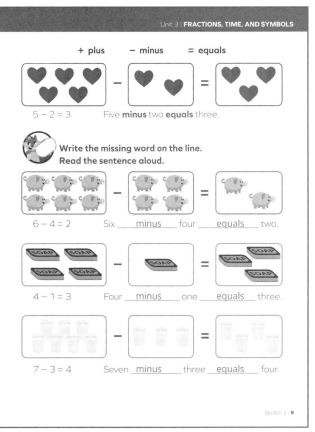

PAGE 10: NUMBER SENTENCES

MATERIALS NEEDED

• pencils

Concept:

reading number sentences

Teacher Goal:

To teach the children to read number sentences in addition and subtraction.

Teaching Page 10:

Turn to page 10. Ask the children to read aloud the number words and the operation symbols at the top of the page. Read the directions with the students. Complete the first line with them so they understand what they are to do. When the page is completed, have the children read each number sentence aloud.

FRACTIONS, TIME, AND SYMBOLS | Unit 3

zero one two three four five six seven eight nine ten

\+ plus   − minus   = equals

Write the symbol or word on the line.
Read the sentence aloud.

1 _+_ 8 = 9	One __plus__ eight equals nine.
_3_ + 2 = 5	__Three__ plus two equals five.
4 + 4 _=_ 8	Four plus four __equals__ eight.
6 _+_ 4 = 10	Six __plus__ four equals ten.
7 + 1 = _8_	Seven plus one equals __eight__ .
7 _−_ 1 = 6	Seven __minus__ one equals six.
9 − 4 _=_ 5	Nine minus four __equals__ five.
9 − 2 = _7_	Nine minus two equals __seven__ .
6 − _1_ = 5	Six minus __one__ equals five.
_9_ − 9 = 0	__Nine__ minus nine equals zero.

10 | Section 2

116

PAGE 11: NUMBER SENTENCES

MATERIALS NEEDED

- pencils
- objects for counting

Concept:

reading number sentences

Teacher Goal:

To teach the children to read the operation symbol for not equal (≠).

Teaching Page 11:

Make a set of *2* objects and a set of *4* objects. Write the fact on the board *2 + 4 = 7*. Ask the children if this is true. Explain to them that we say that "two plus four is not equal to seven." Draw the sign for *not equal* (≠) on the board. Use the example with several other sets until the children understand the concept and sign for *not equal*.

Turn to page 11. Have the students read the words and signs at the top of the page. Go over the sets of brushes and teapots with the children. Have them complete the page.

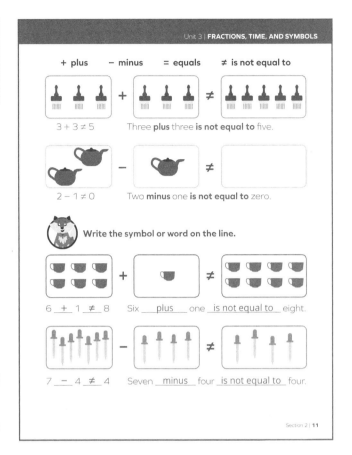

PAGE 12: NUMBER SENTENCES

MATERIALS NEEDED

• pencils

Concept:

reading number sentences

Teacher Goal:

To teach the children to read and write number sentences.

Teaching Page 12:

Turn to page 12 and ask the students to read the words and number symbols at the top of the page. Read the directions and tell the children to point to the first number fact and sentence. Ask them to read the sentence aloud, saying the missing number and the missing word. Have them write the missing symbol and word on the line. Complete several of the number sentences with the students to be sure they understand the directions. Then have them complete the page. Ask the children to say the number sentences aloud as a group or independently when the sentences are complete.

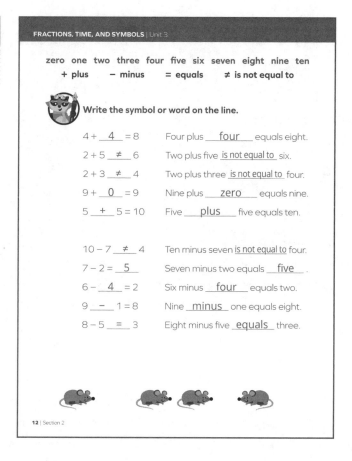

FRACTIONS, TIME, AND SYMBOLS | Unit 3

zero one two three four five six seven eight nine ten
+ plus   − minus   = equals   ≠ is not equal to

Write the symbol or word on the line.

4 + __4__ = 8   Four plus __four__ equals eight.
2 + 5 __≠__ 6   Two plus five _is not equal to_ six.
2 + 3 __≠__ 4   Two plus three _is not equal to_ four.
9 + __0__ = 9   Nine plus __zero__ equals nine.
5 __+__ 5 = 10   Five __plus__ five equals ten.

10 − 7 __≠__ 4   Ten minus seven _is not equal to_ four.
7 − 2 = __5__   Seven minus two equals __five__ .
6 − __4__ = 2   Six minus __four__ equals two.
9 __−__ 1 = 8   Nine __minus__ one equals eight.
8 − 5 __=__ 3   Eight minus five __equals__ three.

12 | Section 2

PAGE 13: NUMBER WORDS

MATERIALS NEEDED

- pencils
- crayons
- ruler

Concepts:

number words to ten, colors

Teacher Goals:

To teach the children to match number symbols and words to 10 and to recognize colors.

Teaching Page 13:

Turn to page 13. Read the instructions aloud with the students. When they understand the instructions, allow them to complete the page independently.

Encourage students to use rulers to draw the matching lines.

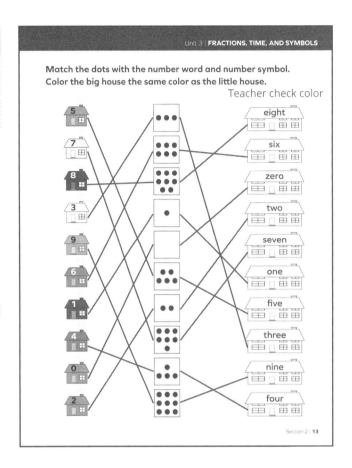

Unit 3 | **FRACTIONS, TIME, AND SYMBOLS**

Match the dots with the number word and number symbol. Color the big house the same color as the little house.

Teacher check color

Section 2 | **13**

PAGE 14: PATTERNS

MATERIALS NEEDED

- pencils
- group of objects (such as pencils, crayons, erasers, or beads)

Concept:

review skills using number patterns

Teacher Goal:

To teach the children to identify the next object in a pattern.

Teaching Page 14:

Review patterns and sequencing with the students. Lay out some patterns using various objects and have the children show the next object or objects in the pattern. Have the students make their own patterns.

Turn to page 14. Read the directions with the children. Tell them that they are to identify each pattern and then show what comes next in the sequence. Allow them to complete the page independently.

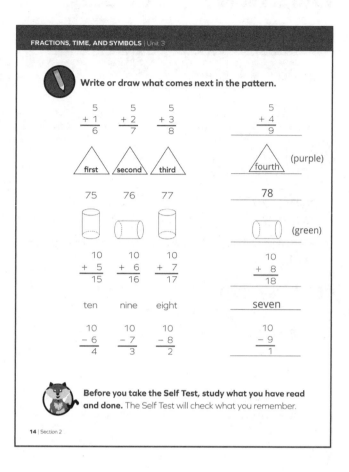

The students should prepare for the Self Test. Ask the students to look over and read the Self Test but they should not write the answers to any questions. After looking over the Self Test the students should go to the beginning of the unit and reread the text and review the answers to the activities up to the Self Test.

The students are to complete the Self Test the next school day. This should be done under regular test conditions without allowing the students to look back. A good idea is to clip the pages together before the test.

PAGE 15: SELF TEST 2

MATERIALS NEEDED

• pencils

Concept:

number sentences

Teacher Goal:

To teach the children to learn to check their progress periodically.

Teaching Page 15:

Turn to page 15. Read the directions to the children. Have the children repeat them after you while running their fingers under the sentence being read. Be sure the children understand what they are to do. You may repeat the directions but give no other help.

Do not have the children check their own work. Check it as soon as you can, and go over it with each child. Show him where he did well and where he needs extra help.

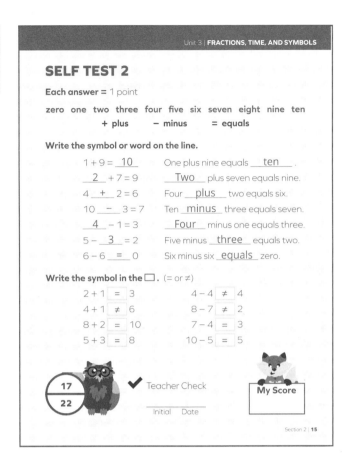

3. FRACTIONS

PAGE 16: FRACTIONS

MATERIALS NEEDED

- pencils
- paper
- scissors
- group of objects that can be divided into two parts
- paper cut into shapes of circle, triangle, rectangle, and square
- glass of water
- ball of clay
- stick
- piece of cloth

Concept:

fraction of one-half

Teacher Goals:

To teach the children to find one-half of an object and to write the fraction ½.

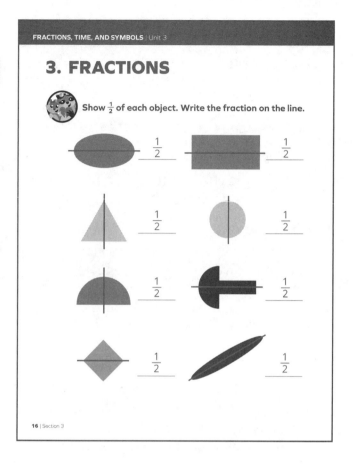

Teaching Page 16:

Introduce the word *one-half* to the students and ask them what the expression means. Begin with the whole glass of water and ask them if they can show *one-half* of a glass of water. Give the students the ball of clay and ask them to give *one-half* of the ball of clay back to you. Show the students the paper circle. Cut the circle into *two* equal parts and hold one part up for the children to see. Write the fraction ½ on the board. Explain to the students that the *2* below the line tells how many parts the circle has been divided into, and the *1* tells how many parts you are talking about. Tell them that this new number is called a *fraction* and it is read *one-half*. Give the students the paper triangle. Have them cut the triangle into *two* equal (emphasize they must be equal) parts. Tell them to draw a line on a piece of paper and write a *2* below the line to illustrate the *two* equal parts. Have them hold up *one* part and then write a *1* above the line to show they are talking about *one* part. Have the students read the fraction. (one-half) Use the remainder of the materials in the same manner.

Turn to page 16 and read the directions aloud. Have the students draw lines to divide each one of the objects into *two* equal parts and then write the fraction.

PAGE 17: FRACTIONS

MATERIALS NEEDED

• pencils
• paper
• groups of objects in even number sets
  (four pencils, ten beads, twelve nickels, six
  crayons, eight blocks)

Concept:

fraction of one-half

Teacher Goals:

To teach the children to find one-half of a set and to write the fraction ½.

Teaching Page 17:

Give the students *four* pencils and ask them to give half of the pencils back to you (2 pencils). Talk to the students again about the meaning of *one-half*. Explain to them that the *four* pencils represent *one* set. Ask them how many parts they divided the set into (2 parts). Have them draw a line on paper and write a *2* below the line to represent the number of parts the set has been

divided into. Have the children hold up the *two* pencils and write a *1* above the line to represent the part of the set they are talking about. Ask the children to read the fraction (one-half). Ask them how many pencils are in *one-half* a set of *four* pencils (2 pencils). Repeat the same procedure with the *ten* beads. Emphasize that even though there are *ten* beads, they represent *one* set. Tell the students to divide the *ten* beads into *two* sets. Write the fraction on paper beginning with the number *2* below the line. Ask how many beads make up *one-half* of a set of *ten*. Continue in this manner with the set of *twelve* nickels, *six* crayons, and *eight* blocks.

Turn to page 17 and read the directions with the students. Emphasize that each picture represents *one* set. Have the students draw a line to divide each set into *two* parts and then write the fraction ½ on the answer line. Ask them to say aloud, "One-half of a set of four boys is two boys" and so on.

PAGE 18: FRACTIONS

MATERIALS NEEDED

- pencils
- paper
- objects to illustrate one-half

Concept:

fraction of one-half

Teacher Goals:

To teach the children to find one-half of a single object or of a set and to write the fraction ½.

Teaching Page 18:

Use the materials to talk about the fraction *one-half* to the children. Show them a single object and have them illustrate *one-half* of the object. Give them a set of objects and have them illustrate *one-half* of one set of objects. Be sure they understand the concept that many objects in a group represents a set of one.

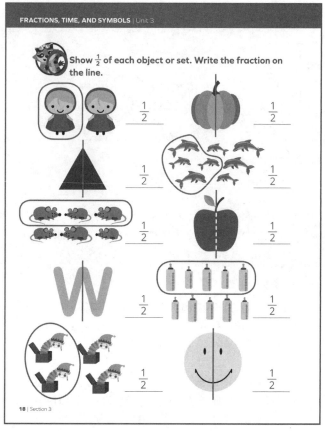

Turn to page 18 and read the directions aloud. Tell the children that they should draw a line for each object or set to show *one-half*. Then they should write the fraction ½ on the answer line.

PAGE 19: FOLLOWING ORAL DIRECTIONS

MATERIALS NEEDED

- pencils
- crayons

Concepts:

fraction of one-half, number order, following oral directions

Teacher Goals:

To teach the children to follow oral directions, to put events in number order, and to find one-half of a single object.

Teaching Page 19:

Turn to page 19. Read the directions at the top of the page and have the students say the numbers aloud. Discuss the pictures on the page so that the children are familiar with each one.

Explain to the students that you will read a story about the pictures to them. They are to put the pictures in number order as they are talked about in the story.

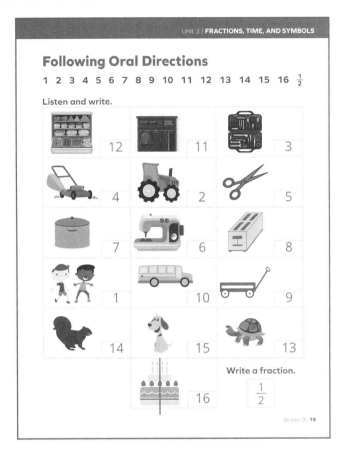

"It is Jason's birthday. Jason and his friend Michael (1) are going to the pet store to buy a birthday present for Jason. Jason's father will stay home and drive his tractor (2). He also needs to use his tools (3) to fix his lawn mower (4). Jason's mother will use her scissors (5) and sewing machine (6) to make a new dress. There is roast beef to cook in the big pan (7). Jason and Michael eat some toast (8) for breakfast and then set off for the store with Jason's wagon (9). They wave to the bus driver as he drives by (10). They pass a clothing store (11) and a vegetable and fruit market (12). Finally, they are at the pet store! There are all kinds of animals. There is a turtle (13), a squirrel (14), and a dog (15). Jason looks and looks but cannot decide. Finally, he selects a pet. Jason's mother has a birthday cake (16) for Jason and Michael when they come home from the store. They are so hungry, they eat the whole cake!"

Have the students complete the page by circling (or drawing) the pet they think Jason picked for his birthday present. Have them draw a line dividing the cake to show that Jason had *one-half* and Michael had *one-half*.

Write the fraction ½ in the box.

PAGE 20: PLACE VALUE

MATERIALS NEEDED

- pencils
- number symbol cards
- objects for counting

Concept:

place value for tens and ones

Teacher Goal:

To teach the children to learn place value for tens and ones.

Teaching Page 20:

Place number symbols cards to represent *26* in front of the students and have them say the number aloud. Ask how many *tens* and how many *ones*. Have the students use objects for counting to make *two* sets of *ten* plus *six ones*.

Reverse the symbols so that the number now reads *62*. Ask the students to say the number aloud. Ask how many *tens* and how many *ones* there are now. Tell the children to illustrate the numbers with objects for counting.

How many?

	tens	ones		tens	ones
42 =	40	+ 2	12 =	10	+ 2
75 =	70	+ 5	95 =	90	+ 5
19 =	10	+ 9	15 =	10	+ 5
38 =	30	+ 8	60 =	60	+ 0

Measure.

_____ 3 inches

_____ 1 inches

_____ 5 inches

Match.

rectangle
circle
triangle
square

Before you take the Self Test, study what you have read and done. The Self Test will check what you remember.

20 | Section 3

Turn to page 20. Point to the number *42*. Ask the students to identify the number in the *tens'* place (4) and the number in the *ones'* place (2). Tell them to point to the *40* and tell them it represents the *4* in the *tens'* place. Have them point to the *2* and tell them this represents how many *ones*. Tell the children to write the values for *tens* (40) and *ones* (2) on the blank lines. Read the other directions on the page and have the children complete the page.

The students should prepare for the Self Test. Ask the students to look over and read the Self Test but they should not write the answers to any questions. After looking over the Self Test the students should go to the beginning of the unit and reread the text and review the answers to the activities up to the Self Test.

The students are to complete the Self Test the next school day. This should be done under regular test conditions without allowing the students to look back. A good idea is to clip the pages together before the test.

PAGE 21: SELF TEST 3

MATERIALS NEEDED

• pencils

Concepts:

fractions, oral directions

Teacher Goal:

To teach the children to learn to check their progress periodically.

Teaching Page 21:

Turn to page 21. Read the directions to the children. Have the children repeat them after you while running their fingers under the sentence being read. Be sure the children understand what they are to do. You may repeat the directions but give no other help.

Read the following story to the students. Tell them to put the story in number order.

"Jack asked his father (1) to guess the kind of pet that he had picked for his birthday present. His father guessed a rabbit (2), a dog (3), a bee (4), or a mouse (5). After Jack showed his father his new pet, Jack and his father divided an apple (6) and each ate one-half."

Do not have the children check their own work. Check it as soon as you can, and go over it with each child. Show him where he did well and where he needs extra help.

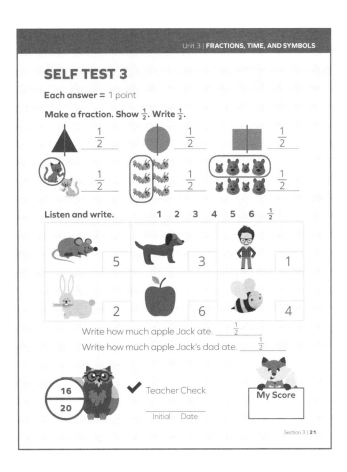

4.   FOLLOWING DIRECTIONS

PAGE 22: FOLLOWING ORAL DIRECTIONS

MATERIALS NEEDED

- pencils
- crayons
- ruler

Concept:

following oral directions

Teacher Goal:

To teach the children to practice math skills following oral directions.

Teaching Page 22:

Turn to page 22. Explain to the children that they should listen to the directions and follow them asking as few questions as possible. Tell them that you will review the page with them when it is completed.

Ask the children to put their fingers on the first apple. You will dictate four problems. They should write the problems vertically in number symbols in the boxes.

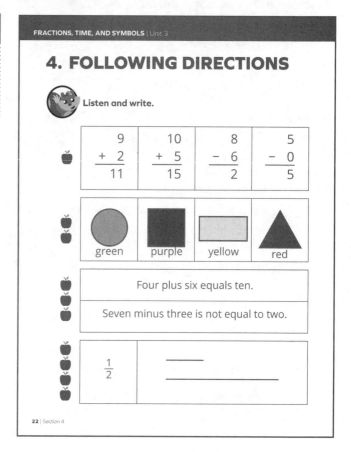

$$
\begin{array}{r} 9 \\ +\ 2 \\ \hline 11 \end{array}
\qquad
\begin{array}{r} 10 \\ +\ 5 \\ \hline 15 \end{array}
\qquad
\begin{array}{r} 8 \\ -\ 6 \\ \hline 2 \end{array}
\qquad
\begin{array}{r} 5 \\ -\ 0 \\ \hline 5 \end{array}
$$

Ask the children to put their fingers on the set of two apples and follow the directions. Draw a *circle* in the *first* box and color it *green*. Draw a *square* in the *second* box and color it *purple*. Draw a *rectangle* in the *third* box and color it *yellow*. Draw a *triangle* in the *fourth* box and color it *red*.

Ask the children to put their fingers on the set of three apples. You will dictate two problems. They should write the number sentences horizontally in number words.

Four plus six equals ten.

Seven minus three is not equal to two.

Ask the children to put their fingers on the set of four apples and follow the directions. Write the fraction ½. Draw lines that measure one inch and three inches.

PAGE 23: COUNT TO 100

MATERIALS NEEDED

- pencils
- chart of numbers
- objects for counting (10's)

Concept:

counting to 100

Teacher Goal:

To teach the children to write numbers to 100.

Teaching Page 23:

Turn to page 23 and read the directions. Have the students complete the page independently. They may use their chart of numbers as necessary. When completed, have them recite the numbers orally.

Tell the children to point to the number *100* on page 23. Have them say "one hundred" aloud. Call attention to the last row of numbers in their chart and have the children point to the numbers in the *tens'* place.

(10 - 20 - 30 - 40 - 50 - 60 - 70 - 80 - 90 - 100)

Show the children that the *10* in *100* is the next number in sequence after the *9* in *90*. Illustrate *100* using *10* objects for counting. Have the students add the number *100* to their chart of numbers.

Unit 3 | FRACTIONS, TIME, AND SYMBOLS

Write the missing numbers. Count to 100.

0	1	2	3	4	5	6	7	8	9
10	11	12	13	14	15	16	17	18	19
20	21	22	23	24	25	26	27	28	29
30	31	32	33	34	35	36	37	38	39
40	41	42	43	44	45	46	47	48	49
50	51	52	53	54	55	56	57	58	59
60	61	62	63	64	65	66	67	68	69
70	71	72	73	74	75	76	77	78	79
80	81	82	83	84	85	86	87	88	89
90	91	92	93	94	95	96	97	98	99
100									

Section 4 | **23**

PAGE 24: CALENDAR

MATERIALS NEEDED

- pencils
- current calendar showing the months of the year

Concept:

reading the calendar

Teacher Goals:

To teach the children to tell the days of the month and to tell the days of the week.

Teaching Page 24:

Spend some time with the students discussing the current calendar. Show the months of the year and that different months have different numbers of days. Have them count the number of days in several months. Point to the days of the week and say them aloud. Ask the students to name the *first* day (Sunday) and the *last* day of the week. (Saturday) Show them that the *first* day of the month may be on any day of the week.

Calendar

May

Sunday	Monday	Tuesday	Wednesday	Thursday	Friday	Saturday
		1	2	3	4	5
6	7	8	9	10	11	12
13	14	15	16	17	18	19
20	21	22	23	24	25	26
27	28	29	30	31		

Name the month of the calendar. ____May____

How many days are in this month? ____31____

How many days are in a week? ____7____

Say the names aloud.

What is the first day of this month?

____Tuesday____   ____May____   ____1____
(day of week)   (month)   (date)

What is the last day of this month?

____Thursday____   ____May____   ____31____
(day of week)   (month)   (date)

24 | Section 4

Turn to page 24. Point to the days of the week and have the students say them aloud. Have the children put their fingers on the number *1* and ask them what is the name of the *first* day of this month. (Tuesday)

Have them fill in the numbers of the month. Read the questions at the bottom of the page aloud with the students and have them write the answers. Keep the current calendar in the classroom. Ask the students every morning to find the day on the calendar. Ask them to name the month, the day of the month, and the day of the week. This should be part of their daily program.

PAGE 25: CALENDAR

MATERIALS NEEDED

- pencils
- current calendar showing the months of the year

Concept:

reading the calendar

Teacher Goals:

To teach the children to tell the months of the year and to tell the number of days in a week and the number of months in a year.

Teaching Page 25:

Turn to page 25. Read the months of the year aloud with the students. Have them look at the current calendar and write the number of days in each month in the box below the name of the month.

Read the rhyme:

Thirty days hath September,
April, June, and November;

All the rest have thirty-one,
excepting February alone.

And that has twenty-eight days clear,
and twenty-nine in each leap year.

Students will find it helpful to learn this rhyme by heart. Have them complete the questions at the bottom of page 25. You may want to explain the exception of February.

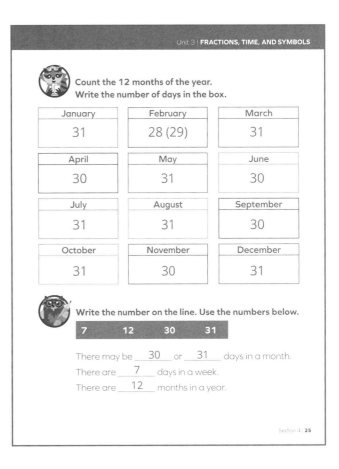

Unit 3 | FRACTIONS, TIME, AND SYMBOLS

Count the 12 months of the year.
Write the number of days in the box.

January	February	March
31	28 (29)	31

April	May	June
30	31	30

July	August	September
31	31	30

October	November	December
31	30	31

Write the number on the line. Use the numbers below.

| 7 | 12 | 30 | 31 |

There may be ___30___ or ___31___ days in a month.
There are ___7___ days in a week.
There are ___12___ months in a year.

Section 4 | 25

PAGE 26: CLOCKS

MATERIALS NEEDED

- pencils
- clock
- scissors
- cardboard
- glue
- brass brad

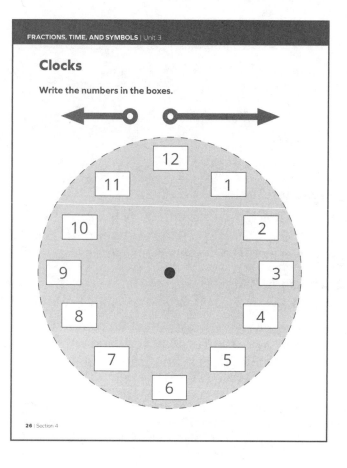

FRACTIONS, TIME, AND SYMBOLS | Unit 3

Clocks

Write the numbers in the boxes.

12
11
1
10
2
9
3
8
4
7
5
6

26 | Section 4

Concept:

reading the clock

Teacher Goal:

To teach the children to tell time to the hour.

Teaching Page 26:

Introduce the clock to the students. Point out the face of the clock to them. Have them identify the big hand and the little hand. Tell them to point to and read the numbers on the clock.

Turn to page 26. Ask the children to identify the face of the clock, the big hand, and the little hand. Read the directions at the top of the page, and tell the students to fill in the numbers. Tell the children that they will make a clock of their own today.

Have them:

1. cut out the clock face (circle) and glue to it to a piece of cardboard,

2. cut out the clock hands and glue them to cardboard,

3. punch a hole in the hands and in the center of the clock, and

4. use a brass brad to hold the hands and clock together.

Be sure the hands are loose enough so the students may move them. This clock should be kept for later lessons. Keep the clock with other materials in the classroom. Ask the children to tell the time on a regular basis.

PAGE 27: CLOCKS

MATERIALS NEEDED

• pencils
• clock made on page 26

Concept:

reading the clock

Teacher Goal:

To teach the children to tell time to the hour.

Teaching Page 27:

Turn to page 27. Ask the children to point to the first clock. Ask them to identify the big hand and the little hand. Have them move the hands on their own clocks to the same positions. Tell them that the little hand tells the hour when the big hand is on the twelve.

Have them say "eight o'clock" aloud and write an *8:00* on the line. Continue in this manner to complete the page.

Be sure the students find the hour on their own clocks for each clock on page 27.

The students should prepare for the Self Test. Ask the students to look over and read the Self Test but they should not write the answers to any questions. After looking over the Self Test the students should go to the beginning of the unit and reread the text and review the answers to the activities up to the Self Test.

The students are to complete the Self Test the next school day. This should be done under regular test conditions without allowing the students to look back. A good idea is to clip the pages together before the test.

Unit 3 | **FRACTIONS, TIME, AND SYMBOLS**

Write the number on the _____ .

8 : 00   12 : 00   1 : 00   4 : 00

5 : 00   10 : 00   7 : 00   2 : 00

1 : 00   12 : 00   6 : 00   11 : 00

3 : 00   9 : 00   2 : 00   4 : 00

Before you take the Self Test, study what you have read and done. The Self Test will check what you remember.

Section 4 | **27**

PAGE 28: SELF TEST 4

MATERIALS NEEDED

• pencils

Concept:

reading a calendar and telling time

Teacher Goal:

To teach the children to learn to check their progress periodically.

Teaching Page 28:

Turn to page 28. Read the directions to the children. Have the children repeat them after you while running their fingers under the sentence being read. Be sure the children understand what they are to do. You may repeat the directions but give no other help.

Do not have the children check their own work. Check it as soon as you can, and go over it with each child. Show him where he did well and where he needs extra help.

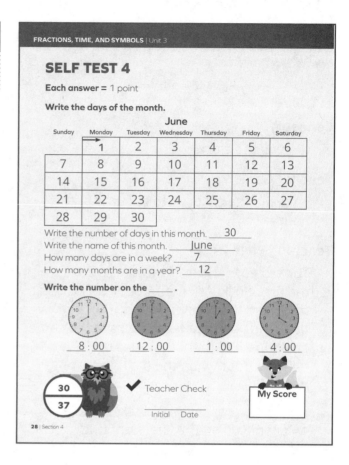

FRACTIONS, TIME, AND SYMBOLS | Unit 3

SELF TEST 4

Each answer = 1 point

Write the days of the month.

June

Sunday	Monday	Tuesday	Wednesday	Thursday	Friday	Saturday
	→ 1	2	3	4	5	6
7	8	9	10	11	12	13
14	15	16	17	18	19	20
21	22	23	24	25	26	27
28	29	30				

Write the number of days in this month. ___30___
Write the name of this month. ___June___
How many days are in a week? ___7___
How many months are in a year? ___12___

Write the number on the _____ .

8 : 00   12 : 00   1 : 00   4 : 00

30
37

✔ Teacher Check

_____ _____
Initial   Date

My Score

28 | Section 4

5.   NUMBER ORDER TO 100

PAGE 29: NUMBER ORDER

MATERIALS NEEDED

• pencils
• chart of numbers
• objects for counting (tens and ones)
  (see LIFEPAC 102, page 20)

Concept:

number order to 50

Teacher Goal:

To teach the children to write numbers in order to 50.

Teaching Page 29:

Introduce the objects for counting to the students. Talk to them about the color strips that represent *10*. Have the students use groups of strips to represent different numbers to *50*. (34 equals three tens' strips and four ones' strips.)

5. NUMBER ORDER TO 100

Write the numbers in order.

10	6	9	8	7
6	7	8	9	10

20	17	19	18	16
16	17	18	19	20

29	30	28	27	31
27	28	29	30	31

42	39	41	40	43
39	40	41	42	43

49	50	52	51	48
48	49	50	51	52

Section 5 : **29**

Continue doing this for several numbers.
Have the students arrange some of these sets of strips in number order. (Example: Make sets for 15, 24, 36, and then arrange them in number order.) Tell the students to locate these numbers on their number chart.

Turn to page 29 and read the directions at the top of the page. Have the children arrange the first group of numbers in number order. Remind them to look at the number in the *tens'* place first and then the number in the *ones'* place. Allow them to complete the page independently.

PAGE 30: NUMBER ORDER

MATERIALS NEEDED

- pencils
- chart of numbers
- objects for counting (tens and ones)
  (see LIFEPAC 102, page 20)

Concept:

number order to 100

Teacher Goal:

To teach the children to write numbers in order to 100.

Teaching Page 30:

Have the students use the counting strips to form numbers between *51* and *100*. (61 equals six tens' strips and one of the ones' strips.) Continue doing this for several numbers. Have the students arrange some of these sets of strips in number order. (Example: Make sets for 57, 72, 83 and then arrange them in number order.) Tell the students to locate these numbers on their number chart.

Write the numbers in order.

57	59	58	60	56
_56_ ,	_57_ ,	_58_ ,	_59_ ,	_60_

69	67	70	66	68
_66_ ,	_67_ ,	_68_ ,	_69_ ,	_70_

75	78	77	79	76
_75_ ,	_76_ ,	_77_ ,	_78_ ,	_79_

86	89	87	88	85
_85_ ,	_86_ ,	_87_ ,	_88_ ,	_89_

96	98	100	99	97
_96_ ,	_97_ ,	_98_ ,	_99_ ,	_100_

30 | Section 5

Turn to page 30 and read the directions at the top of the page. Have the children arrange the first group of numbers in number order. Remind them to look at the number in the *tens'* place first and then the number in the *ones'* place. Allow them to complete the page independently.

PAGE 31: NUMBER WORDS AND NUMBER ORDER

MATERIALS NEEDED

- pencils
- chart of numbers
- objects for counting (tens and ones)

Concept:

number words and number order to 100

Teacher Goals:

To teach the children to read number words to ten and to write the number before and after.

Teaching Page 31:

Use the *tens'* strips and the *ones'* strips to form the number *16*. Ask the students to say the number that comes after *16*. (17) Have them illustrate this by adding *one* strip to the set. Again form the number *16*. Have the students say the number that comes before. Have them illustrate this by taking *one* strip away from the set. Do this exercise with several numbers selecting numbers between *one* and *ninety-nine*.

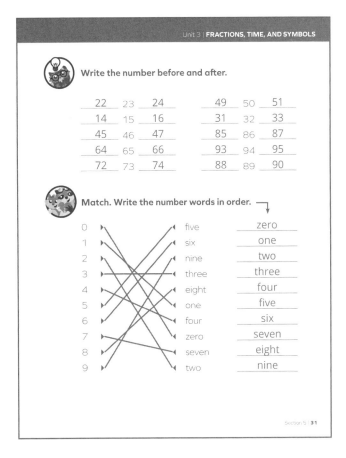

Turn to page 31 and read the directions at the top of the page. Students who have difficulty should use the counting strips or the chart of numbers to find their answers. Ask them to read the directions to the second and third exercises and say the number symbols and the number words aloud. Allow them to complete these exercises independently.

PAGE 32: STORY PROBLEMS

MATERIALS NEEDED

- pencils
- paper
- addition and subtraction fact cards

Concept:

writing answers to story problems

Teacher Goals:

To teach the children to listen to a story problem, to remember details, and to find the answer by writing the problem on paper.

Teaching Page 32:

Review the addition and subtraction fact cards with the students. Dictate the following facts to the students and have them write the facts on paper (vertically using number symbols). Remind the students to use the correct operation symbols (+ and −) and to draw a line between the problem and the answer.

Dictate: (One plus seven equals eight, etc.)

$$\frac{\begin{array}{r}1\\+\ 7\end{array}}{8} \qquad \frac{\begin{array}{r}3\\-\ 1\end{array}}{2} \qquad \frac{\begin{array}{r}8\\-\ 6\end{array}}{2} \qquad \frac{\begin{array}{r}4\\+\ 4\end{array}}{8} \qquad \frac{\begin{array}{r}9\\-\ 5\end{array}}{4} \qquad \frac{\begin{array}{r}2\\+\ 7\end{array}}{9}$$

Turn to page 32 and have the students identify the pictures. Explain to them that they should listen to the story about each set of pictures and write a number fact for the story. Students who have difficulty should begin by illustrating their stories, then write the number fact.

Read:

1. Jenny's mother went to the store. She bought two boxes of cereal and one stalk of celery. How many objects did she have in her bag when she left the store? (2 + 1 = 3)

2. Mrs. Caterpillar was taking her five baby caterpillars for a walk. Two baby caterpillars were tired and went home. How many baby caterpillars were left to take a walk? (5 − 2 = 3)

3. Some of the animals at the zoo were putting on a show. Four of them were singing and two were dancing. How many animals altogether were putting on the show? (4 + 2 = 6)

4. Betty was going to take the train to the city. It was cold outside so she needed warm clothes. There were six buttons on her coat but one was loose and fell off. How many buttons were left on her coat? (6 − 1 = 5)

Story Problems

Listen to the story. Tell how many.

Light Bites box, celery stalk — How many? **3 objects**		$\begin{array}{r}2\\+\ 1\\\hline 3\end{array}$
caterpillar — How many? **3 caterpillars**		$\begin{array}{r}5\\-\ 2\\\hline 3\end{array}$
monkey, alligator, bear, cat — How many? **6 animals**		$\begin{array}{r}4\\+\ 2\\\hline 6\end{array}$
city buildings, coat — How many? **5 buttons**		$\begin{array}{r}6\\-\ 1\\\hline 5\end{array}$

32 | Section 5

PAGE 33: NUMBER ORDER

MATERIALS NEEDED

- pencils
- crayons
- shoe box
- paste/glue
- colored paper

Concept:

number order

Teacher Goals:

To teach the children to write numbers in order and to identify colors.

Teaching Page 33:

Turn to page 33 and read the story. Read the directions with the students. When the students have written the numbers in number order, have them complete the page by drawing something that might be in the treasure chest, or something they would put in a treasure chest. Some children may enjoy making a treasure chest using a shoe box and decorating with crayons or colored paper.

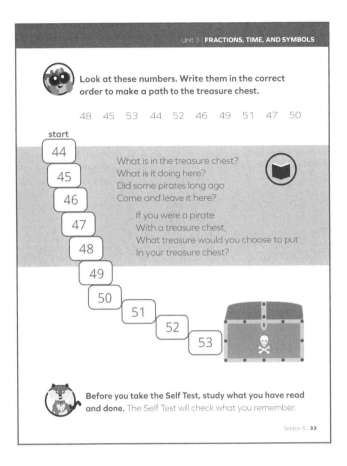

The students should prepare for the Self Test. Ask the students to look over and read the Self Test but they should not write the answers to any questions. After looking over the Self Test the students should go to the beginning of the unit and reread the text and review the answers to the activities up to the Self Test.

The students are to complete the Self Test the next school day. This should be done under regular test conditions without allowing the students to look back. A good idea is to clip the pages together before the test.

PAGE 34: SELF TEST 5

MATERIALS NEEDED

• pencils

Concepts:

number order, story problems

Teacher Goal:

To teach the children to learn to check their progress periodically.

Teaching Page 34:

Turn to page 34. Read the directions to the children. Have the children repeat them after you while running their fingers under the sentence being read. Be sure the children understand what they are to do. You may repeat the directions but give no other help.

Read the story:

Jim and Jack were collecting marbles. Jim had three marbles and Jack had four. How many marbles did they have altogether?

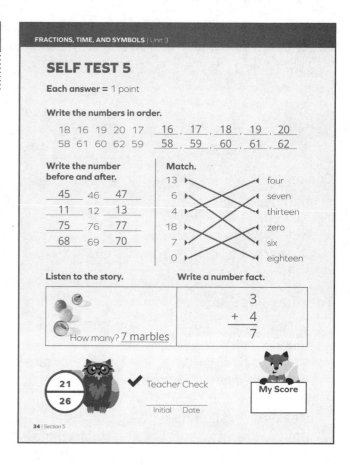

Do not have the children check their own work. Check it as soon as you can, and go over it with each child. Show him where he did well and where he needs extra help.

The students should prepare for the LIFEPAC Test. The students should go to the beginning of the unit and reread the text and review the answers to the activities for the entire unit. Ask the students questions to check their understanding of the unit.

The students are to complete the LIFEPAC Test the next school day. This should be done under regular test conditions without allowing the students to look at the unit.

LIFEPAC TEST 103

MATERIALS NEEDED

• pencils

Concepts:

addition and subtraction, number sentences, signs for = and ≠, writing fractions, reading the calendar, time to the hour, number order, solving oral story problems

Teacher Goal:

To teach the children to learn to check their own progress periodically.

Teaching the LIFEPAC Test:

Administer the test in at least two sessions.

Read all of the directions on each page as the children prepare to do it. Be sure that they understand what they are being asked to do.

LIFEPAC Test page 3

Listen to the story. Write a number fact.

Aaron started the day with 8 marbles. He lost 2 marbles. How many marbles did he have at the end of the day?

Give no help except with directions.

Go over each page with the child as soon as possible after you check it so that he can see where he did well and where he needs more work.

Evaluate the tests and review areas where the children have done poorly. Review the pages and activities that stress the concepts tested.

If necessary, when the children have reviewed sufficiently, administer the Alternate LIFEPAC test. Follow the same procedures as used for the LIFEPAC Test.

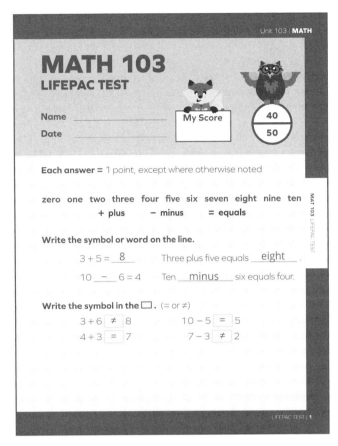

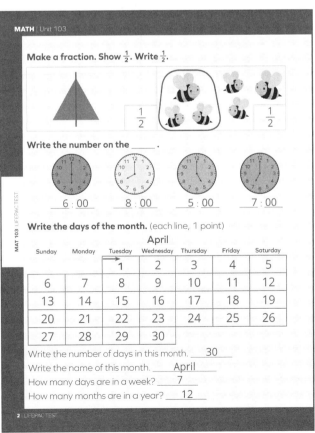

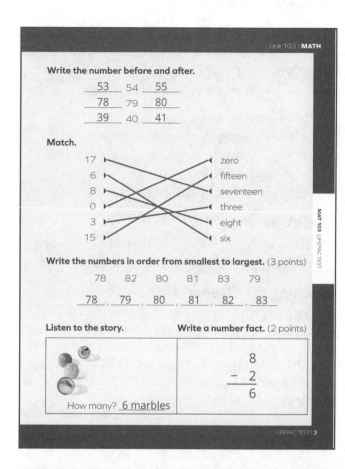

ALTERNATE LIFEPAC TEST 103

MATERIALS NEEDED

• pencils

Concepts:

addition and subtraction, number sentences, signs for = and ≠, writing fractions, reading the calendar, time to the hour, number order, solving oral story problems

Teacher Goal:

To teach the children to learn to check their own progress periodically.

Teaching the Alternate LIFEPAC Test:

Administer the test in at least two sessions.

Read all of the directions on each page as the children prepare to do it. Be sure that they understand what they are being asked to do.

Alternate LIFEPAC Test page 4

Listen to the story. Write a number fact.

There were 9 cookies in the cookie jar. Lisa ate 4 cookies. How many cookies were left in the cookie jar?

Give no help except with directions.

Go over each page with the child as soon as possible after you check it so that he can see where he did well and where he needs more work.

Evaluate the tests and review areas where the children have done poorly. Review the pages and activities that stress the concepts tested.

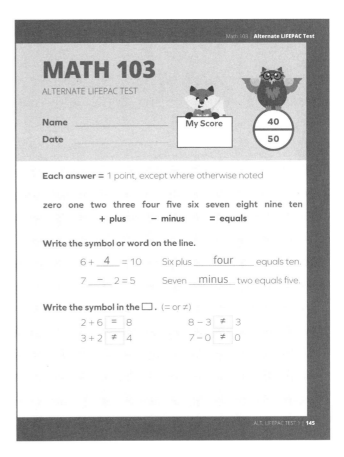

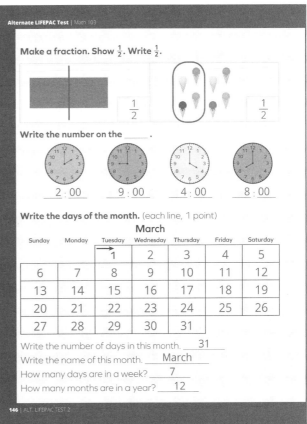

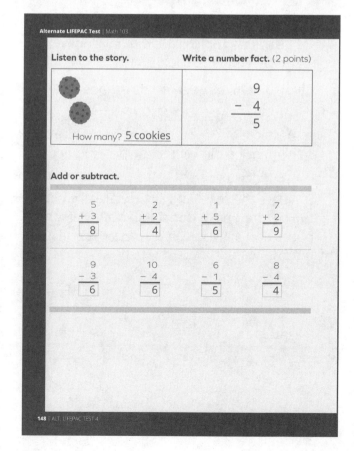

Write the number before and after.

_18_ 19 _20_

_37_ 38 _39_

_69_ 70 _71_

Match.

4 ▸   ◂ six
6 ▸   ◂ two
8 ▸   ◂ four
9 ▸   ◂ zero
0 ▸   ◂ nine
2 ▸   ◂ eight

Write the numbers in order from smallest to largest. (3 points)

21   19   23   20   22   18

_18_ , _19_ , _20_ , _21_ , _22_ , _23_

Listen to the story.   **Write a number fact.** (2 points)

How many? _5 cookies_

$$\begin{array}{r} 9 \\ -\ 4 \\ \hline 5 \end{array}$$

Add or subtract.

$$\begin{array}{r} 5 \\ +\ 3 \\ \hline 8 \end{array} \qquad \begin{array}{r} 2 \\ +\ 2 \\ \hline 4 \end{array} \qquad \begin{array}{r} 1 \\ +\ 5 \\ \hline 6 \end{array} \qquad \begin{array}{r} 7 \\ +\ 2 \\ \hline 9 \end{array}$$

$$\begin{array}{r} 9 \\ -\ 3 \\ \hline 6 \end{array} \qquad \begin{array}{r} 10 \\ -\ 4 \\ \hline 6 \end{array} \qquad \begin{array}{r} 6 \\ -\ 1 \\ \hline 5 \end{array} \qquad \begin{array}{r} 8 \\ -\ 4 \\ \hline 4 \end{array}$$

MATH 103

ALTERNATE LIFEPAC TEST

Name _____

Date _____

My Score

40

50

Each answer = 1 point, except where otherwise noted

zero one two three four five six seven eight nine ten

+ plus   − minus   = equals

Write the symbol or word on the line.

6 + _____ = 10   Six plus _____ equals ten.

7 _____ 2 = 5   Seven _____ two equals five.

Write the symbol in the ▢ . (= or ≠)

2 + 6 ▢ 8   8 − 3 ▢ 3

3 + 2 ▢ 4   7 − 0 ▢ 0

Make a fraction. Show $\frac{1}{2}$. Write $\frac{1}{2}$.

Write the number on the _____ .

_____ : _____   _____ : _____   _____ : _____   _____ : _____

Write the days of the month. (each line, 1 point)

March

Sunday	Monday	Tuesday	Wednesday	Thursday	Friday	Saturday
		1				

Write the number of days in this month. _____

Write the name of this month. _____

How many days are in a week? _____

How many months are in a year? _____

Write the number before and after.

_____ 19 _____

_____ 38 _____

_____ 70 _____

Match.

4 ▶   ◀ six

6 ▶   ◀ two

8 ▶   ◀ four

9 ▶   ◀ zero

0 ▶   ◀ nine

2 ▶   ◀ eight

Write the numbers in order from smallest to largest. (3 points)

21   19   23   20   22   18

_____ , _____ , _____ , _____ , _____ , _____

Listen to the story.

How many? _____

Write a number fact. (2 points)

Add or subtract.

5 + 3	2 + 2	1 + 5	7 + 2

9 − 3	10 − 4	6 − 1	8 − 4

MATH 104

Unit 4: Add to 18, Money, Measurement

ADD TO 18, MONEY, MEASUREMENT
MATH 104

Alpha Omega
PUBLICATIONS

804 N. 2nd Ave. E.
Rock Rapids, IA 51246-1759

© MCMXCVII by Alpha Omega Publications, Inc.
All rights reserved.
LIFEPAC is a registered trademark of Alpha
Omega Publications, Inc.

All trademarks and/or service marks referenced in this
material are the property of their respective owners. Alpha
Omega Publications, Inc. makes no claim of ownership to any
trademarks and/or service marks other than their own and their
affiliates, and makes no claim of affiliation to any companies
whose trademarks may be listed in this material, other than
their own.

Author:
Carol Bauler, B.A.

Editor:
Alan Christopherson, M.S.

Media Credits:
Page 1: © wenchiawang, iStock, Thinkstock;
6: © kazberry, iStock, Thinkstock, **32:** © Volha
Hlinskaya, iStock, Thinkstock; © RobinOlimb,
iStock, Thinkstock.

PAGE 1: ADD TO 18, MONEY, MEASUREMENT

MATERIALS NEEDED

- pencils

Concepts:

purpose of LIFEPAC, objectives

Teacher Goals:

To teach the children to know what is expected of the student in the LIFEPAC and to write first and last names correctly.

Teaching Page 1:

Turn to page 1. Point to the title and read it aloud. Allow time for the children to look through the LIFEPAC. Write the word *Objectives* on the board and have the children find the word on the page. Explain that the objectives tell the things the students will be expected to do in the LIFEPAC. Read each one and have the children repeat them as they run their fingers under the sentence from left to right. Talk about the objectives so that the children will understand what they will be doing. Have each child write his name on the line.

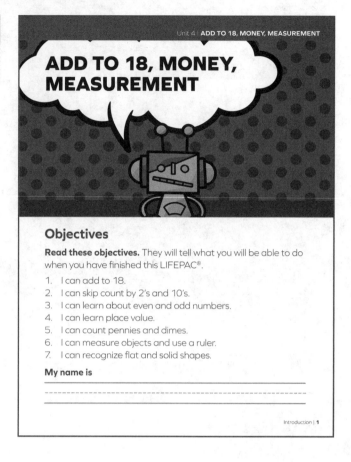

Unit 4 | **ADD TO 18, MONEY, MEASUREMENT**

ADD TO 18, MONEY, MEASUREMENT

Objectives

Read these objectives. They will tell what you will be able to do when you have finished this LIFEPAC®.

1. I can add to 18.
2. I can skip count by 2's and 10's.
3. I can learn about even and odd numbers.
4. I can learn place value.
5. I can count pennies and dimes.
6. I can measure objects and use a ruler.
7. I can recognize flat and solid shapes.

My name is

Introduction | **1**

1. ADDITION FACTS TO 18

PAGES 2 & 3: ADDITION FACTS

MATERIALS NEEDED

• pencils
• addition fact cards

Concept:

addition facts to 18

Teacher Goal:

To teach the children to add facts to 18 using a number line.

Teaching Pages 2 and 3:

Turn to page 2. Ask the children whether they remember counting on the number line. Have them count to eighteen on the number line, pointing to each number as they count. Tell them that they will use the number line to help them learn their new addition facts. Read the directions with the children and have them put their fingers at 6 on the second number line. Tell them to count *three* more. Ask them where they are on the number line now. (9)

Point to the number fact at the bottom of the page. Have the children say aloud, "Six plus three is equal to nine." Remind the children that this is a number sentence. Have the students find this fact in their set of addition fact cards.

Turn to page 3. Read the directions aloud with the children and work with them as they complete the page.

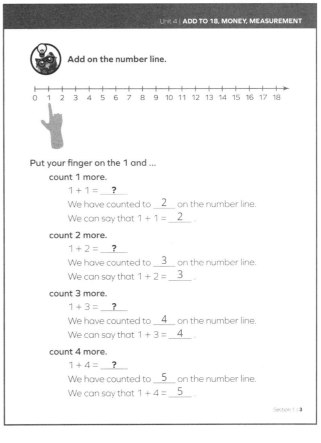

PAGES 4 & 5: ADDITION FACTS

MATERIALS NEEDED

- pencils
- addition fact cards for 1's, 2's and 3's
- new fact cards for 2 + 9, 3 + 8, and 3 + 9
- number symbol cards
- new number symbol cards for 11 and 12

Concept:

addition facts to 18

Teacher Goal:

To teach the children to add facts to 18 using a number line.

Teaching Pages 4 and 5:

Tell the children that they have already learned many of the addition number facts. Tell them that today they will begin working with the remainder of the facts. Explain to them that they should have learned all the facts to *ten* and now they will learn the facts to *eighteen*.

Turn to pages 4 and 5. Ask the children to point to the number line. Tell them that the number line will help them find the answer to the facts until they learn them by heart. Have the children complete the first two lines of facts. Ask them if there is a new fact that they have not learned. (2 + 9) Help them find the answer to the number fact by using the number line. The children should be able to complete the line of *3's* until they reach *3 + 8*. Help them to find the answers to *3 + 8* and *3 + 9* by using the number line. Continue in this manner until pages 4 and 5 are complete. Introduce the new fact cards (2 + 9, 3 + 8, 3 + 9). Play a game of concentration using the fact cards for *2's* and *3's* and the number symbol cards.

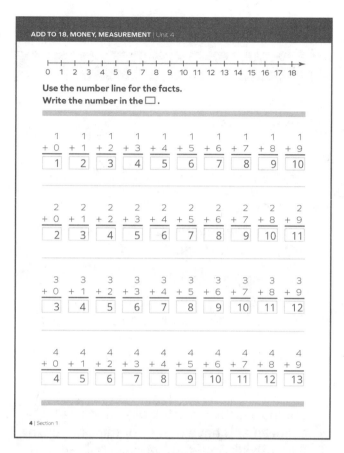

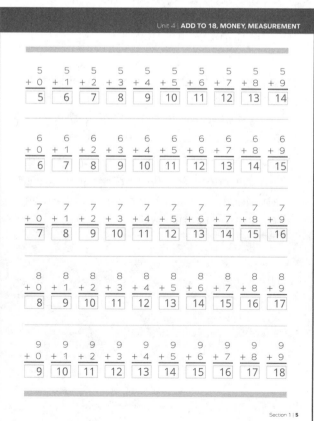

PAGE 6: LISTENING SKILLS, STORY PROBLEMS

MATERIALS NEEDED

- pencils
- addition fact cards for 4's
- new fact cards for 4 + 7, 4 + 8, and 4 + 9
- number symbol cards
- new number symbol card for 13

Concepts:

develop listening skills, solve story problems

Teacher Goals:

To teach the children to write facts and number sentences from dictation and to read and solve story problems.

Teaching Page 6:

Turn to page 6. Tell the children that you are going to dictate four facts to them and that they should write the facts in the boxes (vertically). Remind them to put in the operation sign (+) and to draw a line between the problem and the answer.

Dictate:

$$
\begin{array}{c}
2 \\
+\ 9 \\
\hline
11
\end{array}
\qquad
\begin{array}{c}
3 \\
+\ 8 \\
\hline
11
\end{array}
\qquad
\begin{array}{c}
3 \\
+\ 9 \\
\hline
12
\end{array}
\qquad
\begin{array}{c}
4 \\
+\ 8 \\
\hline
12
\end{array}
$$

Dictate the following number sentences and tell the children to write them in the next two boxes. These sentences should be written in number words horizontally.

Four plus three equals seven.

Five minus two is not equal to six.

Have the children read the stories, write a number fact for each story and write the answer on the line.

Introduce the new fact cards (4 + 7, 4 + 8, 4 + 9) and have the children use objects for counting to illustrate them. Play a game of concentration using the fact cards for *4's* and the number symbol cards.

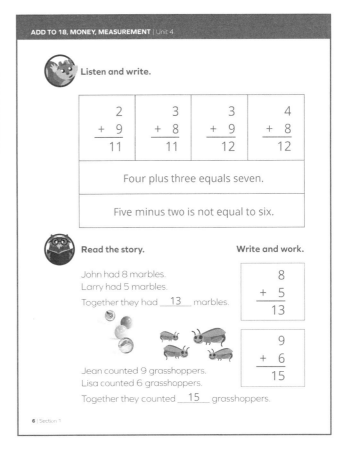

PAGE 7: REVIEWING FACTS

MATERIALS NEEDED

- pencils
- green, blue, red, and brown crayons

Concept:

review addition and subtraction facts

Teacher Goals:

To teach the children to use colors of green, blue, red, and brown and to review addition and subtraction facts.

Teaching Page 7:

Turn to page 7. Read the directions at the top of the page with the children and allow them to complete the page independently.

The students should prepare for the Self Test. Ask the students to look over and read the Self Test but they should not write the answers to any questions. After looking over the Self Test the students should go to the beginning of the unit and reread the text and review the answers to the activities up to the Self Test.

The students are to complete the Self Test the next school day. This should be done under regular test conditions without allowing the students to look back. A good idea is to clip the pages together before the test.

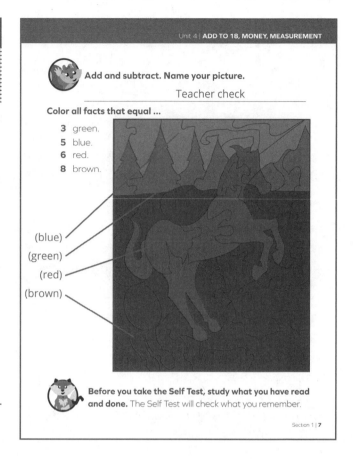

Unit 4 | **ADD TO 18, MONEY, MEASUREMENT**

Add and subtract. Name your picture.

Teacher check

Color all facts that equal ...

3 green.
5 blue.
6 red.
8 brown.

(blue)
(green)
(red)
(brown)

Before you take the Self Test, study what you have read and done. The Self Test will check what you remember.

Section 1 | **7**

PAGE 8: SELF TEST 1

MATERIALS NEEDED

• pencils

Concept:

adding numbers to 18

Teacher Goal:

To teach the children to learn to check their progress periodically.

Teaching Page 8:

Turn to page 8. Read the directions to the children. Have the children repeat them after you while running their fingers under the sentence being read. Be sure the children understand what they are to do. You may repeat the directions but give no other help. Children may use the number line to find their answers to the first set of questions. Dictate five number facts to the students to write in the boxes. The first four are in number symbols and the last is in number words.

Dictate:

$$
\begin{array}{cccc}
2 & 3 & 4 & 4 \\
+\ 9 & +\ 8 & +\ 7 & +\ 8 \\
\hline
11 & 11 & 11 & 12
\end{array}
$$

Five plus four equals nine.

Do not have the children check their own work. Check it as soon as you can, and go over it with each child. Show him where he did well and where he needs extra help.

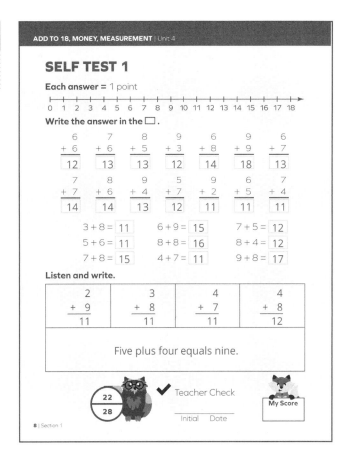

ADD TO 18, MONEY, MEASUREMENT | Unit 4

SELF TEST 1

Each answer = 1 point

0 1 2 3 4 5 6 7 8 9 10 11 12 13 14 15 16 17 18

Write the answer in the ☐.

$$
\begin{array}{ccccccc}
6 & 7 & 8 & 6 & 9 & 9 & 6 \\
+\ 6 & +\ 6 & +\ 5 & +\ 3 & +\ 8 & +\ 9 & +\ 7 \\
\hline
12 & 13 & 13 & 12 & 14 & 18 & 13
\end{array}
$$

$$
\begin{array}{ccccccc}
7 & 8 & 9 & 5 & 9 & 6 & 7 \\
+\ 7 & +\ 6 & +\ 4 & +\ 7 & +\ 2 & +\ 5 & +\ 4 \\
\hline
14 & 14 & 13 & 12 & 11 & 11 & 11
\end{array}
$$

3 + 8 = 11	6 + 9 = 15	7 + 5 = 12
5 + 6 = 11	8 + 8 = 16	8 + 4 = 12
7 + 8 = 15	4 + 7 = 11	9 + 8 = 17

Listen and write.

2	3	4	4
+ 9	+ 8	+ 7	+ 8
11	11	11	12

Five plus four equals nine.

22 / 28

✔ Teacher Check

Initial   Date

My Score

8 | Section 1

2. SKIP COUNTING

PAGE 9: COUNT BY 2

MATERIALS NEEDED

• pencils
• chart of numbers (LIFEPAC 101, page 7)
• objects for counting

Concept:

counting by 2's

Teacher Goal:

To teach the children to count by 2's.

Teaching Page 9:

Have the students look at their chart of numbers as they count to *ten*. Have them clap on the *even* numbers as they say 1 - 2 - 3 - 4 - 5 - 6 - 7 - 8 - 9 - 10.

Explain to the children that sometimes we want to count quickly and so we do not count every number. We count certain numbers and that is called *skip counting*. Tell them that today they will learn to skip count by 2. Ask the children to point to the chart again and this time say 2 - 4 - 6 - 8 - 10. Have the children make *five* sets with *two* objects in each set. Have them take *one* set and count 2; take the *second* set, count 4; the *third* set, count 6; the *fourth* set, count 8; and the *fifth* set, count 10. Have them count the whole set and ask if they have counted correctly. (yes)

Turn to page 9. Read the instructions on the page. Tell the children to count aloud, first reading the words and then the numbers. Have the children illustrate the ducks using objects for counting.

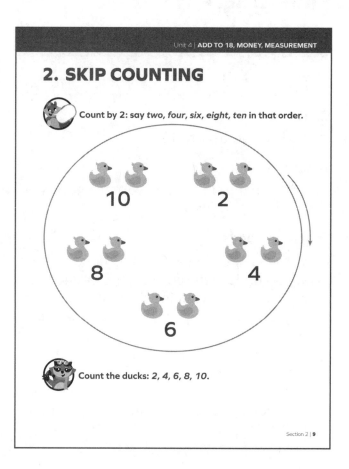

2. SKIP COUNTING

Count by 2: say *two, four, six, eight, ten* in that order.

10   2

8   4

6

Count the ducks: *2, 4, 6, 8, 10.*

Section 2 | **9**

PAGE 10: COUNT BY 2'S

MATERIALS NEEDED

- pencils
- chart of numbers
- addition fact cards through 4's

Concept:

counting by 2's

Teacher Goal:

To teach the children to count by 2's.

Teaching Page 10:

Have the children use their chart of numbers to count to *twenty* clapping their hands on the *even* numbers. Have them point to the chart of numbers and skip count by *2's* to *twenty*. Emphasize to them that they are grouping the numbers by *2's*.

Turn to page 10. Ask the children how the objects are shown in each box. (grouped by 2's) Have the children write the answer in the box counting by *2's*. Children who need to may use the chart of numbers.

Review the addition fact cards through *4's*. Go through them once in order and then scramble them.

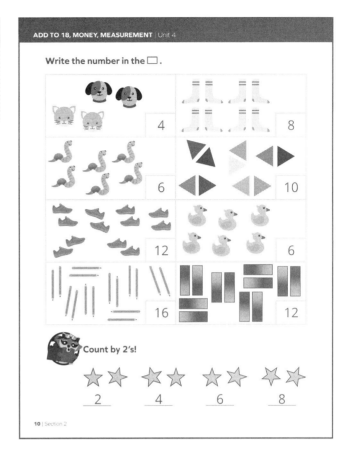

PAGE 11: ODD AND EVEN

MATERIALS NEEDED

- pencils
- chart of numbers
- addition fact cards through 4's

Concept:

learning odd and even numbers

Teacher Goal:

To teach the children to count odd and even numbers.

Teaching Page 11:

Introduce the words *odd* and *even* to the students. Explain to them that we use these expressions to describe certain numbers. Have the children use their chart of numbers to skip count by *2's* to *twenty*. Tell the children that the numbers they said aloud are called *even* numbers and the numbers they skipped are called *odd* numbers.

Turn to page 11. Tell the children to point to the words *odd* and *even* at the top of the chart and to say them aloud. Ask the children to say the *odd* numbers together and the *even* numbers together.

Read the instructions at the top of the page with the children. Have them read the sentences and write the answers.

Review addition facts through *4's*.

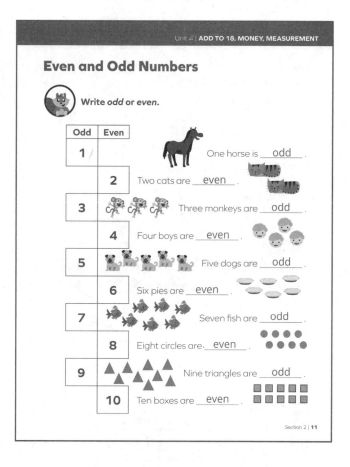

PAGE 12: COUNT BY 10'S

MATERIALS NEEDED

- pencils
- chart of numbers
- addition fact cards for 5's (5 + 0 through 5 + 5)
- number symbol cards

Concept:

counting by 10's

Teacher Goal:

To teach the children to count by 10's.

Teaching Page 12:

Tell the children to use their objects for counting to make *ten* sets of *10* objects. Have them start with a set of *10* and point to the *10* on the chart of numbers. Have them add a set of *ten* to the set they already have and ask "how many?" Have them point to the *20* on their number chart.

Continue in this manner adding *ten* more each time, until the children reach *100*. Have the students look at their chart of numbers as they say *10 - 20 - 30 - 40 - 50 - 60 - 70 - 80 - 90 - 100*. Explain to the children that this is called *skip counting* by *10*. Have the children point to the chart again and skip count by *10* again.

Turn to page 12. Read the instructions on the page. Tell the children to count the number of stars in the first circle and write the number on the line. Ask them if they think there are the same number of stars in each circle. Ask them if they think they can skip count the number of stars on the page. (yes, by 10's) Have them skip count and say the answer aloud. Have the students begin counting the stars individually until they reach *60*. As they tire of counting, ask them if they are glad they have learned to count by *10's*. Tell them that this is the reason that people have learned to skip count.

Play a game of concentration with fact cards for *5's* and number symbol cards.

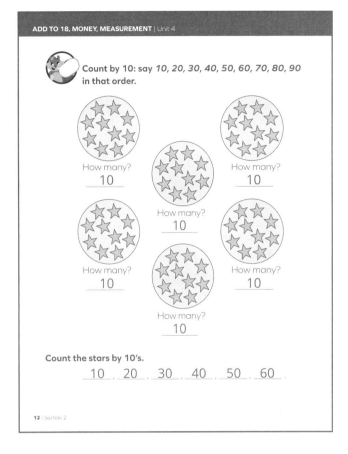

PAGE 13: COUNT BY 10'S

MATERIALS NEEDED

- pencils
- chart of numbers
- objects for counting
- fact cards for 5's
- new fact cards for 5 + 6, 5 + 7, 5 + 8, and 5 + 9
- new number symbol card 14

Concept:

counting by 10's

Teacher Goal:

To teach the children to count by 10's.

Teaching Page 13:

Using the chart of numbers, have the students skip count by *2* to *twenty*; then have them skip count by *10* to *one hundred*. Have them put the chart away and ask them to skip count the same way from memory. Students having difficulty with the concept should go back over the exercise using objects for counting working with sets of *2's* or *10's* as needed.

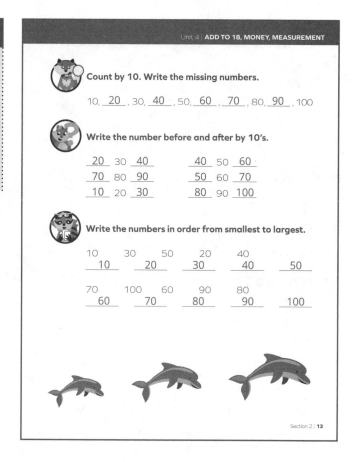

Unit 4 | ADD TO 18, MONEY, MEASUREMENT

Count by 10. Write the missing numbers.

10, _20_, 30, _40_, 50, _60_, 70, 80, _90_, 100

Write the number before and after by 10's.

_20_ 30 _40_   _40_ 50 _60_
_70_ 80 _90_   _50_ 60 _70_
_10_ 20 _30_   _80_ 90 _100_

Write the numbers in order from smallest to largest.

10   30   50   20   40
_10_   _20_   _30_   _40_   _50_

70   100   60   90   80
_60_   _70_   _80_   _90_   _100_

Section 2 | **13**

Turn to page 13. Read the directions and have the students complete the first exercise. Read the directions for the second exercise. Be sure the children understand this is an exercise in skip counting.

For example:

The answer to the first problem is *20 30 40*, not *29 30 31*.

Read the instructions to the final exercise and have the students complete the page.

Review the fact cards with the students for *5's*. Introduce the new fact cards for *5 + 6, 5 + 7, 5 + 8*, and *5 + 9*, and have the students illustrate each one using objects for counting.

PAGE 14: NUMBER ORDER

MATERIALS NEEDED

• pencils
• chart of numbers
• crayons

Concept:

number order

Teacher Goals:

To teach the children to count by 1's and to write numbers in number order.

Teaching Page 14:

Tell the children that it is still important that they know how to count by *1's*. Have them use their chart of numbers to count aloud from *1* to *100*.

Turn to page 14 and read the rhyme aloud with the children. Ask them if they think the cowboy is riding a horse. Read the directions on the page aloud and have the children complete the page independently. Let them color the picture when the exercises are completed.

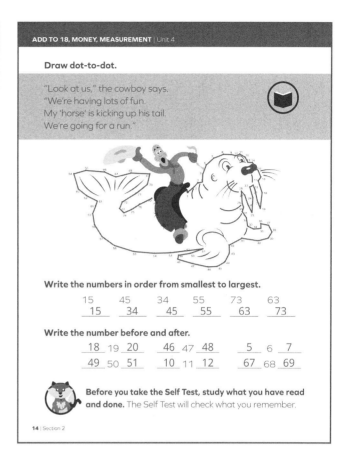

The students should prepare for the Self Test. Ask the students to look over and read the Self Test but they should not write the answers to any questions. After looking over the Self Test the students should go to the beginning of the unit and reread the text and review the answers to the activities up to the Self Test.

The students are to complete the Self Test the next school day. This should be done under regular test conditions without allowing the students to look back. A good idea is to clip the pages together before the test.

PAGE 15: SELF TEST 2

MATERIALS NEEDED

- pencils
- chart of numbers
- fact cards for 5's (5 + 0 through 5 + 9)
- number symbol cards

Concept:

counting by 1's, 2's, and 10's

Teacher Goal:

To teach the children to learn to check their progress periodically.

Teaching Page 15:

Turn to page 15. Read the directions to the children. Have the children repeat them after you while running their fingers under the sentence being read. Be sure the children understand what they are to do. You may repeat the directions but give no other help. Children may use their chart of numbers for this page.

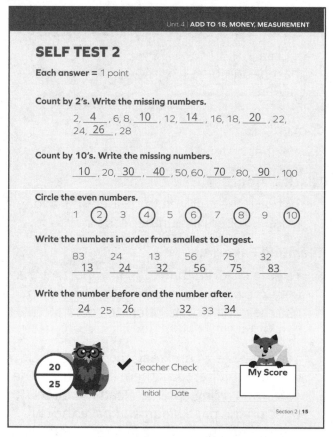

Do not have the children check their own work. Check it as soon as you can, and go over it with each child. Show him where he did well and where he needs extra help.

When the page is completed, scramble the fact cards for *5's*. Have the students select the correct number symbol card to go with each fact card.

3. NUMBERS AS WORDS

PAGES 16 & 17: NUMBER WORDS TO TWENTY

MATERIALS NEEDED

- pencils
- subtraction fact cards 1's through 5's (LIFEPAC 101, page 19 and LIFEPAC 102, page 7)
- objects for counting

Concept:

number words to twenty

Teacher Goal:

To teach the children to read number words to twenty.

Teaching Pages 16 and 17:

Turn to page 16. Ask the children to read aloud the number symbols and the number words in the first section. Have them read the number symbols in the second section. Ask them if they think they know what the number words are in the second section. Have the students point to the number symbol *11* and then point to the number word *eleven*. Have them look at the word and say the word aloud.

Continue in this manner having the students say each number word from *eleven* through *twenty*.

Point to the next two sections. Tell the children to match the number symbols and the number words. Tell them they may look at the first two sections to complete the exercises.

Turn to page 17 and read the directions aloud. Tell the children they may look at

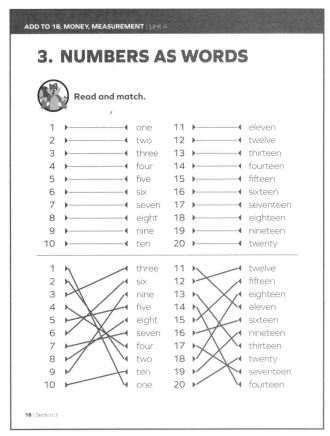

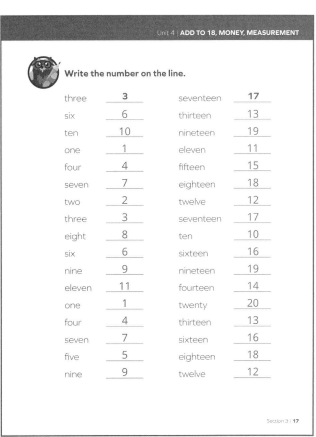

page 16 to find the numbers. When they have completed the page have them point to each one of the number words and say it aloud.

Review the subtraction fact cards through *5's*. First do them in order and then scramble them. Those that are missed should be reviewed using objects for counting.

PAGE 18: NUMBER WORDS TO TWENTY

MATERIALS NEEDED

- pencils
- subtraction fact cards 6's through 9's (LIFEPAC 101, page 19 and LIFEPAC 102, page 7)
- objects for counting

Concept:

number words to twenty

Teacher Goal:

To teach the children to read number words to twenty.

Teaching Page 18:

Turn to page 18. Have the students read the words at the top of the page. Read the directions aloud with the students. Tell them that this time they are to write the number words. Students have learned to count in number order. When they realize the words are in number order they should be able to find the correct words for this exercise easily. Remind them to sit straight, to hold their pencils correctly, and to keep their letters on the line.

Review the subtraction fact cards for *6's* through *9's*. First do them in order and then scramble them. Those that are missed should be reviewed using objects for counting.

ADD TO 18, MONEY, MEASUREMENT | Unit 4

one	six	eleven	sixteen
two	seven	twelve	seventeen
three	eight	thirteen	eighteen
four	nine	fourteen	nineteen
five	ten	fifteen	twenty

Write the word on the line.

4	four	15	fifteen
7	seven	19	nineteen
5	five	12	twelve
10	ten	14	fourteen
3	three	16	sixteen
6	six	11	eleven
1	one	20	twenty
2	two	13	thirteen
9	nine	18	eighteen
8	eight	17	seventeen

18 | Section 3

PAGES 19 & 20: PLACE VALUE

MATERIALS NEEDED

- pencils
- objects for counting (tens and ones)

Concept:

place values to 100

Teacher Goal:

To teach the children to write numbers in tens and ones.

Teaching Pages 19 and 20:

Review with the students the meaning of the objects for counting that are marked for *tens* and for *ones*. Have the students illustrate seven or eight numbers using the objects. For example: *48* would be *4* of the *tens'* strips and *8* of the *ones'* strips. Reverse the procedure by placing sets of objects in front of the students and ask them to tell the number. For example: *6 tens'* strips and *3 ones'* strips would be *63*. (Students should be able to count the tens' strips by skip counting by 10.)

Turn to page 19. Point to the first illustration and count the blocks with the students. Show them which one is the same as their *tens'* strips and which ones are the same as their *ones'* strips. Point to the line that says *tens* and the number written on the line. (10) Point to the line that says *ones* and the number that is written on the line. (4) Follow the same procedure for the next two illustrations and have the students write the numbers on the lines. (20, 6 and 30, 8) Point out to the children how to write *10 + 4 = 14*. Have the students write the missing values in the four sets.

Turn to page 20. Have the children illustrate the first ten problems using objects for counting (tens and ones) as they write the number on the line. Allow them to complete the page independently. When they have finished, have them read the numbers aloud that they have written.

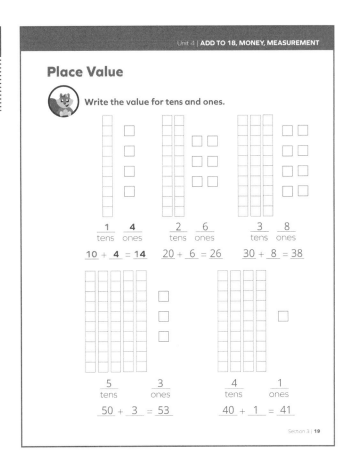

Place Value

Write the value for tens and ones.

1	4
tens	ones

$10 + 4 = 14$

2	6
tens	ones

$20 + 6 = 26$

3	8
tens	ones

$30 + 8 = 38$

5	3
tens	ones

$50 + 3 = 53$

4	1
tens	ones

$40 + 1 = 41$

Section 3 | **19**

Write the number on the line.

$10 + 6 = 16$ $80 + 2 = 82$

$10 + 9 = 19$ $70 + 6 = 76$

$20 + 3 = 23$ $60 + 5 = 65$

$50 + 6 = 56$ $80 + 7 = 87$

$30 + 1 = 31$ $90 + 2 = 92$

$10 + 4 = 14$ $80 + 9 = 89$

$30 + 8 = 38$ $90 + 3 = 93$

$20 + 2 = 22$ $70 + 3 = 73$

$10 + 8 = 18$ $60 + 8 = 68$

$40 + 6 = 46$ $90 + 9 = 99$

$20 + 9 = 29$ $60 + 2 = 62$

$30 + 7 = 37$ $80 + 1 = 81$

$40 + 1 = 41$ $70 + 5 = 75$

$50 + 2 = 52$ $90 + 1 = 91$

$30 + 3 = 33$ $80 + 5 = 85$

$50 + 9 = 59$ $60 + 9 = 69$

20 | Section 3

PAGE 21: PLACE VALUE

MATERIALS NEEDED

• pencils

Concept:

place values to 100

Teacher Goal:

To teach the children to write numbers as tens and ones.

Teaching Page 21:

Turn to page 21 and point to the illustration. Have the children count how many *tens* and say the value of the *tens* on the line. (30) Have the children count the *ones* and say the value of the *ones* on the line. (6) Point to the illustration and say together, "36 is equal to 3 in tens plus 6 in ones." Point to the first number in the exercise (32) and ask the students which number is in the *tens'* place (3) and which number is in the *ones'* place. (2) Tell the children to fill in the value of the *tens* and the value of the *ones*. Ask them to say aloud, "32 is equal to 30 plus 2." Have the children complete the page. Monitor their work closely to be sure they understand the process.

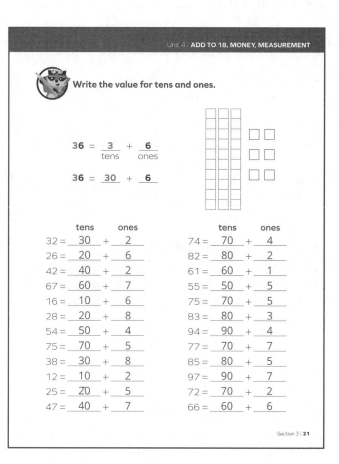

PAGE 22: EQUAL AND NOT EQUAL

MATERIALS NEEDED

- pencils
- objects for counting (ones)

Concept:

identifying equal and not equal

Teacher Goals:

To teach the children to tell number sentences that are equal and not equal and to recognize and write the symbols for equal (=), not equal (≠), plus (+), and minus (−).

Teaching Page 22:

Review the meaning of *equal* and *not equal* with the students. Write the symbols (=, ≠) on the board and ask the students to identify each one. Review the meaning of number sentences.

Turn to page 22. Tell the children that this is a page of number sentences. They must complete the first group of number sentences by writing the sign for *equal* (=) or *not equal* (≠) on the line and the second group by writing *plus* (+) or *minus* (−) on the line. Have the children read the number words at the top of the page. Point out that one column of number sentences is in number symbols and the second column is in number words. Students may use objects for counting if necessary to complete the number facts.

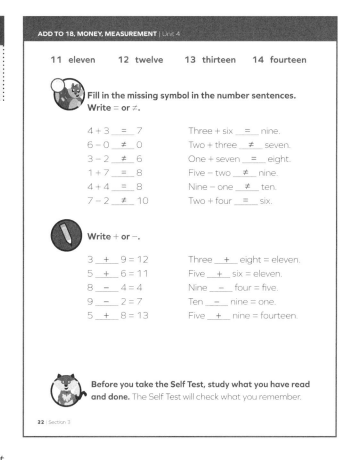

The students should prepare for the Self Test. Ask the students to look over and read the Self Test but they should not write the answers to any questions. After looking over the Self Test the students should go to the beginning of the unit and reread the text and review the answers to the activities up to the Self Test.

The students are to complete the Self Test the next school day. This should be done under regular test conditions without allowing the students to look back. A good idea is to clip the pages together before the test.

PAGE 23: SELF TEST 3

MATERIALS NEEDED

• pencils

Concepts:

expanding numbers, =, ≠

Teacher Goal:

To teach the children to learn to check their progress periodically.

Teaching Page 23:

Turn to page 23. Read the directions to the children. Have the children repeat them after you while running their fingers under the sentence being read. Be sure the children understand what they are to do. You may repeat the directions but give no other help.

Do not have the children check their own work. Check it as soon as you can, and go over it with each child. Show him where he did well and where he needs extra help.

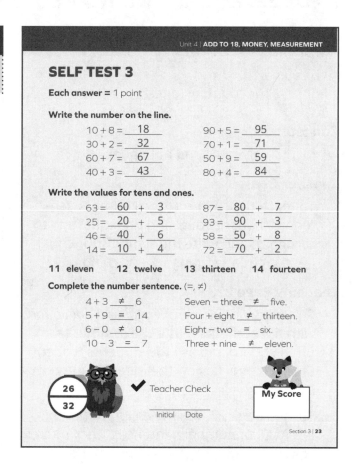

Unit 4 | **ADD TO 18, MONEY, MEASUREMENT**

SELF TEST 3

Each answer = 1 point

Write the number on the line.

$10 + 8 =$ __18__ $90 + 5 =$ __95__
$30 + 2 =$ __32__ $70 + 1 =$ __71__
$60 + 7 =$ __67__ $50 + 9 =$ __59__
$40 + 3 =$ __43__ $80 + 4 =$ __84__

Write the values for tens and ones.

$63 =$ __60__ + __3__ $87 =$ __80__ + __7__
$25 =$ __20__ + __5__ $93 =$ __90__ + __3__
$46 =$ __40__ + __6__ $58 =$ __50__ + __8__
$14 =$ __10__ + __4__ $72 =$ __70__ + __2__

11 eleven   12 twelve   13 thirteen   14 fourteen

Complete the number sentence. (=, ≠)

$4 + 3$ __≠__ 6 Seven − three __≠__ five.
$5 + 9$ __=__ 14 Four + eight __≠__ thirteen.
$6 − 0$ __≠__ 0 Eight − two __=__ six.
$10 − 3$ __=__ 7 Three + nine __≠__ eleven.

26
32

✔ Teacher Check

_____
Initial   Date

My Score

Section 3 | **23**

4.   PENNIES AND DIMES
PAGE 24: COUNTING MONEY

MATERIALS NEEDED

- pencils
- objects for counting (ones and tens)
- pennies and dimes
- beans
- buttons

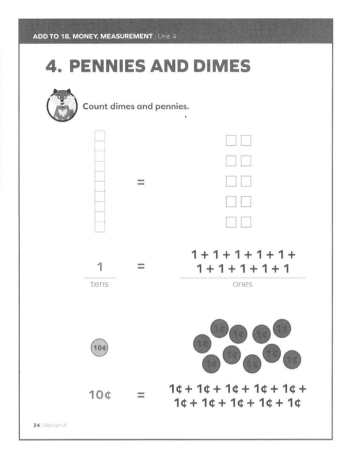

Concept:

identifying pennies and dimes

Teacher Goals:

To teach the children to recognize pennies and dimes and to count how many pennies are equal to one dime.

Teaching Page 24:

Explain to the children that years ago when people wanted something, they would trade an item for it, but they found that they did not always have something to trade. (You might give some examples.) Now we give money in exchange for goods or services. Some of our money is in coins. Two of the coins that we use are pennies and dimes. (Show the pennies and dimes to the children.)

Give the children a group of *twenty beans* and tell them that they are worth *seven pennies*. Have the children count out *seven pennies*. Give the children a group of *five buttons* and tell them that they are worth *sixteen pennies*. Have the children count *sixteen pennies*. Explain to the children that we have *dimes* to make counting money easier. Use the objects for counting (tens and ones) to illustrate several numbers.

For example:

68 = 6 tens + 8 ones.

Have the children illustrate *68¢* using *6 dimes* and *8 pennies*.

Turn to page 24. Remind the children the illustrations are like the strips they use for counting. The first illustration stands for *tens* and the second illustration shows *ones*. Point to the picture of the dimes and pennies. Have the students count how many pennies are equal to one dime.

PAGE 25: COUNTING MONEY

MATERIALS NEEDED

• pencils
• pennies and dimes

Concept:

identifying pennies and dimes

Teacher Goal:

To teach the children to count how many dimes and how many pennies are in a number.

Teaching Page 25:

Turn to page 25 and read the directions at the top of the page. Have the students complete this part of the exercise.

Point to the second exercise. Have the children write how many *tens* and how many *ones*. (10, 5) Tell them to write how many dimes (1) and how many pennies. (5) Tell the children to use their dimes and pennies to count 15¢. Have the children complete this exercise. Be sure they count out the money each time. In the next exercise, the children should write how many cents *6 dimes* and *2 pennies* are equal to. (60, 2; or 62¢) Have the children use the pennies and dimes to illustrate these numbers also.

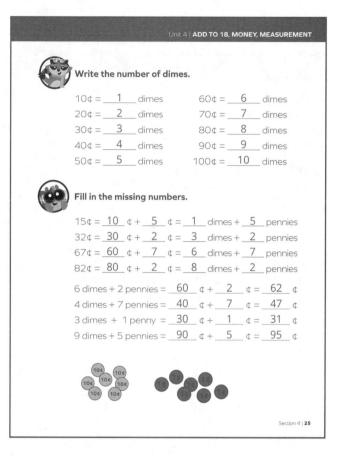

Write the number of dimes.

10¢ = __1__ dimes   60¢ = __6__ dimes
20¢ = __2__ dimes   70¢ = __7__ dimes
30¢ = __3__ dimes   80¢ = __8__ dimes
40¢ = __4__ dimes   90¢ = __9__ dimes
50¢ = __5__ dimes   100¢ = __10__ dimes

Fill in the missing numbers.

15¢ = __10__ ¢ + __5__ ¢ = __1__ dimes + __5__ pennies
32¢ = __30__ ¢ + __2__ ¢ = __3__ dimes + __2__ pennies
67¢ = __60__ ¢ + __7__ ¢ = __6__ dimes + __7__ pennies
82¢ = __80__ ¢ + __2__ ¢ = __8__ dimes + __2__ pennies

6 dimes + 2 pennies = __60__ ¢ + __2__ ¢ = __62__ ¢
4 dimes + 7 pennies = __40__ ¢ + __7__ ¢ = __47__ ¢
3 dimes + 1 penny = __30__ ¢ + __1__ ¢ = __31__ ¢
9 dimes + 5 pennies = __90__ ¢ + __5__ ¢ = __95__ ¢

Section 4 | **25**

PAGE 26: COUNTING MONEY

MATERIALS NEEDED

- pencils
- pennies and dimes
- addition fact cards
- number symbol cards

Concept:

identifying pennies and dimes

Teacher Goal:

To teach the children to count how many dimes and how many pennies are in a number.

Teaching Page 26:

Turn to page 26 and read the instruction at the top of the page. Explain to the students that Jenny is going to the store. The problem is to find out whether she has enough money. Read the first paragraph and see that each student has *eight* dimes and *seven* pennies.

Does Jenny Have Enough Money?

Jenny has some money to spend.
She wants to buy a few things at the store.
Jenny has 87 cents. She has 8 dimes and 7 pennies.

Jenny wants to buy an apple.
The apple costs 22 cents.
The apple costs __2__ dimes and __2__ pennies.
Now Jenny has __6__ dimes and __5__ pennies.
She has __65__ cents.

Jenny wants to buy a soda to drink.
The soda costs 30 cents.
The soda costs __3__ dimes and __0__ pennies.
Now Jenny has __3__ dimes and __5__ pennies.
She has __35__ cents.

Jenny wants to buy two cookies.
The cookies cost 24 cents.
The cookies cost __2__ dimes and __4__ pennies.
Now Jenny has __1__ dimes and __1__ pennies.
She has __11__ cents.

Jenny wants to buy some gum.
The gum costs 11 cents.
The gum costs __1__ dimes and __1__ pennies.
Now Jenny has __0__ dimes and __0__ pennies.
She has __0__ cents.

Does Jenny have enough money? __yes__

How much money does she have left? __0__

26 | Section 4

Read each paragraph with the children.
Have them write the correct answers on the lines. Tell them to take away or put aside the amount of money that has been spent before going on to the next paragraph.

apple:	2 dimes	2 pennies
soda:	3 dimes	
cookie:	2 dimes	4 pennies
gum:	1 dime	1 penny

Review the new addition fact cards with the students. (2 + 9, 3 + 8, 3 + 9, 4 + 7, 4 + 8, 4 + 9, 5 + 6, 5 + 7, 5 + 8, and 5 + 9) Use these fact cards and number symbol cards to play a game of concentration.

PAGE 27: STORY PROBLEMS

MATERIALS NEEDED

- pencils
- pennies and dimes
- new addition fact cards for 6 + 5, 6 + 6, 6 + 7, 6 + 8, and 6 + 9
- number symbol cards
- new number symbol card for 15
- objects for counting

Concept:

story problems with money

Teacher Goal:

To teach the children to solve story problems with money.

Teaching Page 27:

Turn to page 27. Tell the children that these are story problems and that the answers will be in dimes, pennies, or cents. Be sure the children have play money to help them solve the problems. Read the second set of directions. The students should be able to read the stories and complete this page independently.

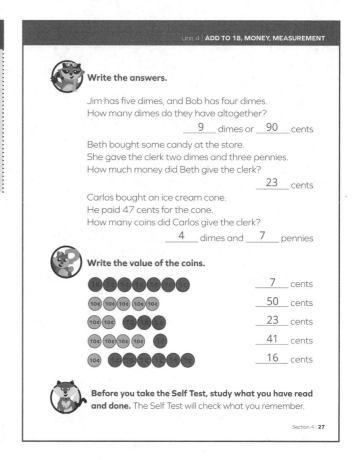

When they have finished, have them use objects for counting to illustrate sets for *6 + 5* through *6 + 9*. Introduce the new fact cards. Have the students match the new fact cards and the fact cards reviewed after the lesson for page 26 with the number symbol cards for *11, 12, 13, 14,* and *15*.

The students should prepare for the Self Test. Ask the students to look over and read the Self Test but they should not write the answers to any questions. After looking over the Self Test the students should go to the beginning of the unit and reread the text and review the answers to the activities up to the Self Test.

The students are to complete the Self Test the next school day. This should be done under regular test conditions without allowing the students to look back. A good idea is to clip the pages together before the test.

PAGE 28: SELF TEST 4

MATERIALS NEEDED

• pencils
• pennies and dimes

Concept:

counting money

Teacher Goal:

To teach the children to learn to check their progress periodically.

Teaching Page 28:

Turn to page 28. Read the directions to the children. Have the children repeat them after you while running their fingers under the sentence being read. Be sure the children understand what they are to do. You may repeat the directions but give no other help. Children may use the pennies and dimes to help illustrate their work.

Do not have the children check their own work. Check it as soon as you can, and go over it with each child. Show him where he did well and where he needs extra help.

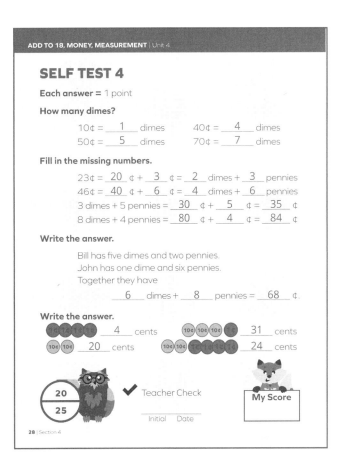

5. SHAPES AND MEASUREMENTS

PAGE 29: FLAT SHAPES

MATERIALS NEEDED

- pencils
- paper
- green, orange, purple, and blue crayons
- fact cards for 6's

Concept:

recognizing flat shapes

Teacher Goal:

To teach the children to recognize flat shapes of circle, triangle, square, and rectangle.

Teaching Page 29:

Draw an illustration of a *circle*, a *triangle*, a *square* and a *rectangle* on the board.

Tell the children to point out in this order:

1. the three-sided figure
2. the figure with four equal sides
3. the round figure
4. the figure with opposite sides equal

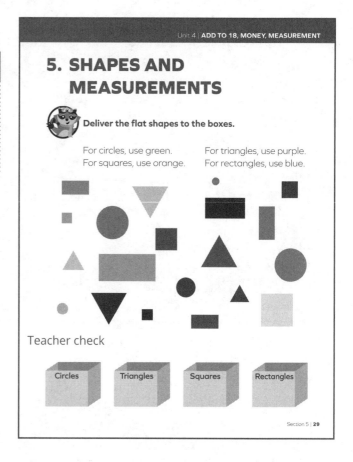

Write the names of each figure on the board and have the students match the name of the figure with the illustration. Ask the children how we describe these figures. (flat—no inside or outside) Have the children say the names of the figures aloud.

Turn to page 29 and read the directions. Point to the boxes at the bottom of the page and have the students read the label on each box. Tell the children to use the color crayon shown in the directions to draw a line from each figure to the box.

Review the new fact cards for *6's*.

PAGE 30: SOLID SHAPES

MATERIALS NEEDED

• pencils
• reference the items listed in LIFEPAC 104, page 30; collect as many as possible

Concept:

recognizing solid shapes

Teacher Goal:

To teach the children to recognize solid shapes of sphere, cylinder, cube, cone, and rectangular solid.

Teaching Page 30:

Have the children go through the items assembled for this lesson and have them identify each one. Discuss how these items are different from the ones they worked with on page 29. (solid, not flat)

Turn to page 30. Have the children say the names of the shapes aloud and then match the names to the shapes. Correct this part of the student page before proceeding to the next exercise. Tell the children to select one of the objects in the list from the group of objects that have been collected. (drinking glass) (If you cannot collect all of the objects, use as many as possible.) Tell the children to find its shape at the top of the page and then write *a*, *b*, *c*, *d*, or *e* on the line. When all the objects have been identified, tell the students to count how many of each kind of shape and write the number on the lines at the bottom of the page. Be sure that you use the names of the shapes and have the students say the shapes as much as possible to help them learn and remember. Save the materials from this page for the lesson on page 31.

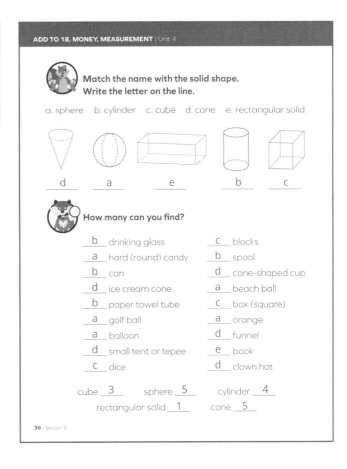

PAGE 31: WEIGHT

MATERIALS NEEDED

- pencils
- list of items from lesson on page 30
- additional items for weighing
- scale that will weigh ounces and/or pounds—a scale for food (dieters) or letters (mailing) that will measure ounces

Concept:

estimating weight

Teacher Goal:

To teach the children to weigh solid shapes.

Teaching Page 31:

Ask the children to arrange the items in order from lightest weight to heaviest weight. Explain to them that they are arranging the items by *estimating* their weight. Allow the students enough time to do this well. Let them compare the weights of different items. Have them start with an empty drinking glass and compare to a full glass.

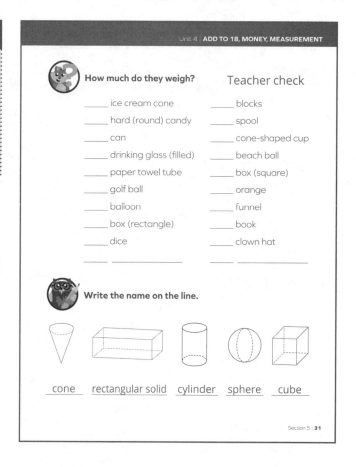

Turn to page 31. Explain to the children how to read the scale that you have for the lesson. Talk to them about ounces and pounds (16 ounces = 1 pound).

If you have collected some items not shown on the page, write the names of the items on the blank lines. Tell the children that they should measure the items using a scale. Weigh each of the collected items on the scale and write the amount on the line. Compare the weights to the order in which the students placed the objects. Discuss where they were right and where they were wrong. Have them write the names of the solid shapes on the lines.

PAGE 32: MEASURING INCHES

MATERIALS NEEDED

- pencils
- addition fact cards for 4's through 6's
- paper
- ruler made in LIFEPAC 101, page 25

Concept:

measuring in inches

Teacher Goal:

To teach the children to use their rulers to measure inches.

Teaching Page 32:

Begin today's lesson with a review of addition fact cards *4's* through *6's*.

Draw several lines on paper (in inches from one to six) and have the students measure each line. Ask the students to say the length of the line aloud. Be sure they use the word *inches* in giving their answer.

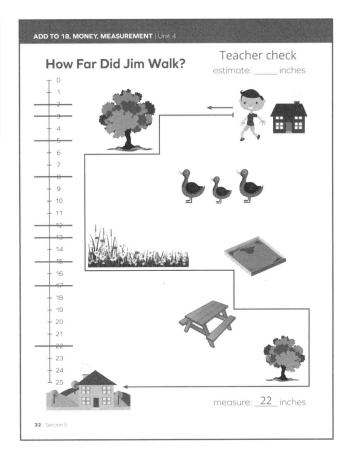

Turn to page 32. Read the directions at the top of the page. Point out the number line.

Have the children count from *zero* to *twenty-five*, pointing to the line as they count. Tell the children that Jim is going to his friend Raymond's house so that they can play catch with Jim's new ball. There are a lot of things to see as Jim walks along, trees and flowers, a sand box in the park, and a picnic table to take a rest. Jim's path is laid out in inches on the paper. First, have the students estimate how far Jim walked (in inches) and write the number on the line at the top of the page. (Do not be concerned if students give exaggerated answers.) Then have them measure the distance. Tell them to measure each line and to keep adding on the number line for each measurement they take.

When finished, the number line should tell the students how far Jim walked. Write the number on the line at the bottom of the page. Have the students compare their estimations and their answers.

PAGE 33: REVIEWING FACTS

MATERIALS NEEDED

• pencils

Concept:

review addition and subtraction facts

Teacher Goal:

To teach the children to review addition and subtraction facts.

Teaching Page 33:

Turn to page 33. Read the directions at the top of the page with the children and allow them to complete the page independently. Be sure they are reading the signs (+ and −) carefully.

The students should prepare for the Self Test. Ask the students to look over and read the Self Test but they should not write the answers to any questions. After looking over the Self Test the students should go to the beginning of the unit and reread the text and review the answers to the activities up to the Self Test.

The students are to complete the Self Test the next school day. This should be done under regular test conditions without allowing the students to look back. A good idea is to clip the pages together before the test.

Unit 4 | **ADD TO 18, MONEY, MEASUREMENT**

0 1 2 3 4 5 6 7 8 9 10 11 12 13 14 15 16 17 18

Add or subtract. Put the answer in the ☐ .

3 + 5 **8**	6 + 8 **14**	9 + 7 **16**	5 + 4 **9**	3 + 2 **5**	6 + 1 **7**
9 − 4 **5**	8 − 3 **5**	6 − 0 **6**	5 − 2 **3**	7 − 4 **3**	8 − 1 **7**
2 + 5 **7**	6 + 9 **15**	8 + 4 **12**	2 + 9 **11**	7 + 3 **10**	5 + 6 **11**
10 − 4 **6**	5 − 1 **4**	7 − 6 **1**	2 − 1 **1**	9 − 6 **3**	6 − 4 **2**
3 + 8 **11**	6 + 7 **13**	5 + 9 **14**	4 + 3 **7**	6 + 6 **12**	2 + 5 **7**
7 − 1 **6**	10 − 5 **5**	3 − 1 **2**	8 − 4 **4**	3 − 2 **1**	8 − 7 **1**

Before you take the Self Test, study what you have read and done. The Self Test will check what you remember.

Section 5 | **33**

PAGE 34: SELF TEST 5

MATERIALS NEEDED

- pencils

Concepts:

shapes, measurements

Teacher Goal:

To teach the children to learn to check their progress periodically.

Teaching Page 34:

Turn to page 34. Read the directions to the children. Have the children repeat them after you while running their fingers under the sentence being read. Be sure the children understand what they are to do. You may repeat the directions but give no other help.

Do not have the children check their own work. Check it as soon as you can, and go over it with each child. Show him where he did well and where he needs extra help.

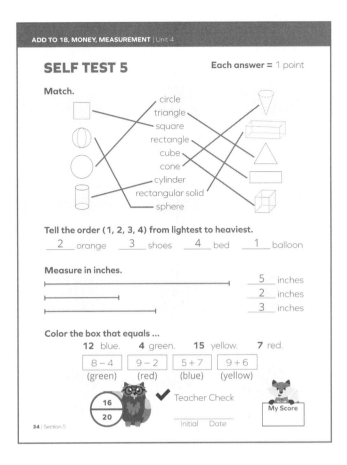

The students should prepare for the LIFEPAC Test. The students should go to the beginning of the unit and reread the text and review the answers to the activities for the entire unit. Ask the students questions to check their understanding of the unit.

The students are to complete the LIFEPAC Test the next school day. This should be done under regular test conditions without allowing the students to look at the unit.

LIFEPAC TEST 104

MATERIALS NEEDED

- pencils
- crayons
- pennies

Concepts:

addition to 18, skip count by 2's and 10's, expanded numbers, pennies and dimes, flat and solid shapes, comparing objects by weight, number words

Teacher Goal:

To teach the children to learn to check their own progress periodically.

Teaching the LIFEPAC Test:

Administer the test in at least two sessions.

Read all of the directions on each page as the children prepare to do it. Be sure that they understand what they are being asked to do.

LIFEPAC Test page 3

Listen and write.

$$\begin{array}{r} 5 \\ +\ 7 \\ \hline 12 \end{array}$$ *Six plus eight equals fourteen.*

Give no help except with directions.

Go over each page with the child as soon as possible after you check it so that he can see where he did well and where he needs more work.

Evaluate the tests and review areas where the children have done poorly. Review the pages and activities that stress the concepts tested.

If necessary, when the children have reviewed sufficiently, administer the Alternate LIFEPAC test. Follow the same procedures as used for the LIFEPAC Test.

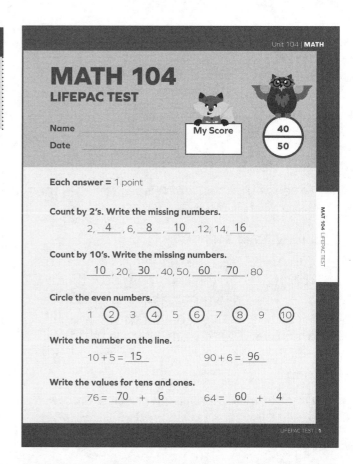

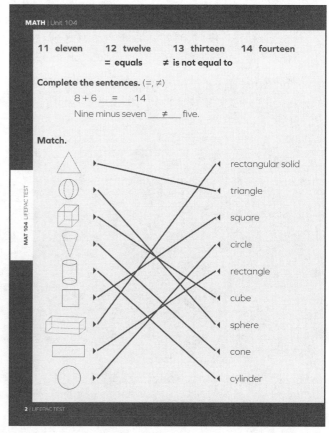

Tell the order (1, 2, 3, 4) from lightest to heaviest.

__3__ brick   __4__ car   __1__ feather   __2__ apple

Fill in the missing numbers.

57¢ = __50__ ¢ + __7__ ¢ = __5__ dimes + __7__ pennies

7 dimes + 4 pennies = __70__ ¢ + __4__ ¢ = __74__ ¢

Listen and write.

$$\begin{array}{r} 5 \\ +\ 7 \\ \hline 12 \end{array}$$

Listen and write.

Six plus eight equals fourteen.

MAT 104 LIFEPAC TEST

LIFEPAC TEST | **3**

MATH | Unit 104

Measure.

__5__ inches

Add.

$$\begin{array}{r} 7 \\ +\ 6 \\ \hline 13 \end{array} \qquad \begin{array}{r} 5 \\ +\ 9 \\ \hline 14 \end{array} \qquad \begin{array}{r} 8 \\ +\ 4 \\ \hline 12 \end{array}$$

$$\begin{array}{r} 6 \\ +\ 8 \\ \hline 14 \end{array} \qquad \begin{array}{r} 9 \\ +\ 9 \\ \hline 18 \end{array} \qquad \begin{array}{r} 4 \\ +\ 7 \\ \hline 11 \end{array}$$

MAT 104 LIFEPAC TEST

4 | LIFEPAC TEST

181

ALTERNATE LIFEPAC TEST 104

MATERIALS NEEDED

- pencils
- crayons
- pennies

Concepts:

addition to 18, skip count by 2's and 10's, expanded numbers, pennies and dimes, flat and solid shapes, comparing objects by weight, number words

Teacher Goal:

To teach the children to learn to check their own progress periodically.

Teaching the Alternate LIFEPAC Test:

Administer the test in at least two sessions.

Read all of the directions on each page as the children prepare to do it. Be sure that they understand what they are being asked to do.

Alternate LIFEPAC Test page 3

Listen and write.

$$\begin{array}{r} 9 \\ + \ 3 \\ \hline 12 \end{array}$$ *Seven plus six equals thirteen.*

Give no help except with directions.

Go over each page with the child as soon as possible after you check it so that he can see where he did well and where he needs more work.

Evaluate the tests and review areas where the children have done poorly. Review the pages and activities that stress the concepts tested.

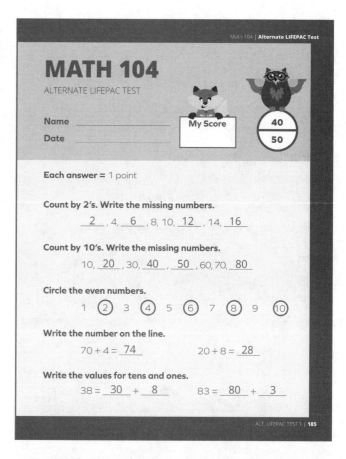

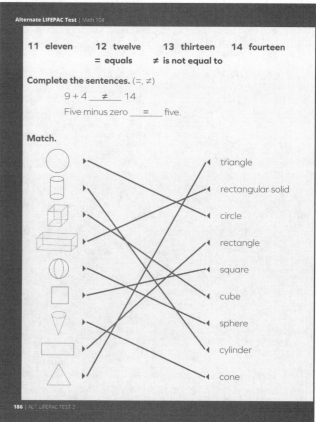

Tell the order (1, 2, 3, 4) from lightest to heaviest.

__3__ puppy __1__ pencil __2__ pear __4__ zebra

Fill in the missing numbers.

34¢ = __30__ ¢ + __4__ ¢ = __3__ dimes + __4__ pennies

8 dimes + 3 pennies = __80__ ¢ + __3__ ¢ = __83__ ¢

Listen and write.

$$\begin{array}{r} 9 \\ + 3 \\ \hline 12 \end{array}$$

Listen and write.

Seven plus six equals thirteen.

Measure.

|—————————————| __4__ inches

Add.

$$\begin{array}{r} 9 \\ + 5 \\ \hline 14 \end{array} \qquad \begin{array}{r} 8 \\ + 8 \\ \hline 16 \end{array} \qquad \begin{array}{r} 4 \\ + 7 \\ \hline 11 \end{array}$$

$$\begin{array}{r} 3 \\ + 9 \\ \hline 12 \end{array} \qquad \begin{array}{r} 8 \\ + 5 \\ \hline 13 \end{array} \qquad \begin{array}{r} 7 \\ + 8 \\ \hline 15 \end{array}$$

MATH 104

ALTERNATE LIFEPAC TEST

Name _____

Date _____

My Score

40
50

Each answer = 1 point

Count by 2's. Write the missing numbers.

_____ , 4, _____ , 8, 10, _____ , 14, _____

Count by 10's. Write the missing numbers.

10, _____ , 30, _____ , _____ , 60, 70, _____

Circle the even numbers.

1   2   3   4   5   6   7   8   9   10

Write the number on the line.

70 + 4 = _____   20 + 8 = _____

Write the values for tens and ones.

38 = _____ + _____   83 = _____ + _____

11 eleven   **12 twelve**   **13 thirteen**   **14 fourteen**

= equals   **≠ is not equal to**

Complete the sentences. (=, ≠)

9 + 4 _____ 14

Five minus zero _____ five.

Match.

 ▶   ◀ triangle

 ▶   ◀ rectangular solid

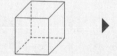 ▶   ◀ circle

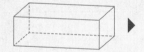 ▶   ◀ rectangle

 ▶   ◀ square

 ▶   ◀ cube

 ▶   ◀ sphere

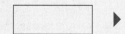 ▶   ◀ cylinder

 ▶   ◀ cone

Tell the order (1, 2, 3, 4) from lightest to heaviest.

_____ puppy   _____ pencil   _____ pear   _____ zebra

Fill in the missing numbers.

34¢ = _____ ¢ + _____ ¢ = _____ dimes + _____ pennies

8 dimes + 3 pennies = _____ ¢ + _____ ¢ = _____ ¢

Listen and write.

Listen and write.

Measure.

_____ inches

Add.

```
   9          8          4
 + 5        + 8        + 7
 ____       ____       ____
```

```
   3          8          7
 + 9        + 5        + 8
 ____       ____       ____
```

MATH 105

Unit 5: Column Addition and Estimation

COLUMN ADDITION AND ESTIMATION
MATH 105

Alpha Omega
PUBLICATIONS

804 N. 2nd Ave. E.
Rock Rapids, IA 51246-1759

Author:
Carol Bauler, B.A.

Editor:
Alan Christopherson, M.S.

Media Credits:
Page 1: © wenchiawang, iStock, Thinkstock;
23: © Valha Hlinskaya, iStock, Thinkstock;
32: © Ashva73, iStock, Thinkstock.

i

PAGE 1: COLUMN ADDITION AND ESTIMATION

MATERIALS NEEDED

• pencils

Concepts:

purpose of LIFEPAC, objectives

Teacher Goals:

To teach the children to know what is expected of the student in the LIFEPAC and to write first and last names correctly.

Teaching Page 1:

Turn to page 1. Point to the title and read it aloud. Allow time for the children to look through the LIFEPAC. Write the word *Objectives* on the board and have the children find the word on the page. Explain that the objectives tell the things the students will be expected to do in the LIFEPAC. Read each one and have the children repeat them as they run their fingers under the sentence from left to right. Talk about the objectives so that the children will understand what they will be doing. Have each child write his name on the line.

Unit 5 | COLUMN ADDITION AND ESTIMATION

COLUMN ADDITION AND ESTIMATION

Objectives

Read these objectives. They will tell what you will be able to do when you have finished this LIFEPAC®.

1. I can add three numbers in columns.
2. I can write fractions for $\frac{1}{2}$ and $\frac{1}{4}$.
3. I can tell time to the half-hour.
4. I can say ordinal numbers.
5. I can make a chart.
6. I can estimate an answer.

My name is

Introduction | **1**

1. COLUMNAR ADDITION

PAGES 2 & 3: ADD TO 18

MATERIALS NEEDED

- pencils
- new fact cards for 7 + 4, 7 + 5, 7 + 6, 7 + 7, 7 + 8, and 7 + 9
- number symbol card 16
- objects for counting

Concept:

add numbers to 18

Teacher Goal:

To teach the children to add numbers to 18 using a number line.

Teaching Pages 2 and 3:

Turn to page 2. Review the number line with the students. Complete the first three facts on page 2 with the students. Have them use the number line on each of these problems. Allow the students to complete the page independently. Some students may require additional help. Encourage them to use the number line.

Turn to page 3. Review the page with the students. Ask them to point out the facts that they should know and those facts for which they will need to use the number line. (Students should know the facts through the 6's or be familiar with them.) Select some fact cards at random and review them with the students. Tell them to complete page 3 using the number line when necessary. Have them illustrate the facts for 7's using objects for counting.

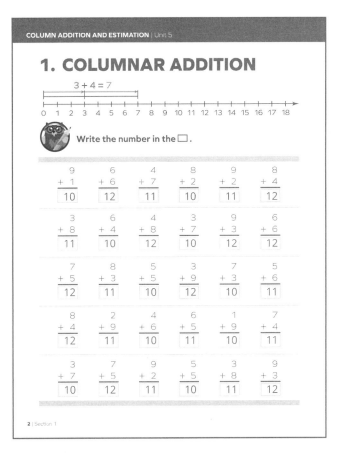

PAGE 4: SUBTRACT TO 10

MATERIALS NEEDED

- pencils
- subtraction fact cards

Concept:

subtract numbers to 10

Teacher Goal:

To teach the children to subtract numbers to 10 using a number line.

Teaching Page 4:

Select some subtraction fact cards at random and review with the students. A complete review should be done at the end of this lesson for students having difficulty with subtraction facts introduced to this point of the LIFEPACs.

Turn to page 4. Point to the number line at the top of the page and explain to the children that we can subtract as well as add on the number line. Ask the children to say the fact aloud at the top of the page. (8 − 3 = 5)

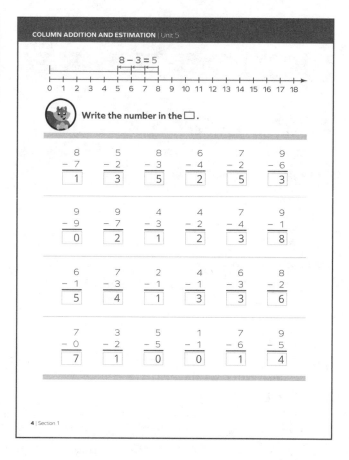

Tell them to point to the *8* on the number line and then count *3* moving from right to left. Explain to them that this is *minus* because they are moving from a *larger* number to a *smaller* number. Have the children illustrate the first three number facts on the number line and then let them complete the page independently.

PAGES 5 & 6: COLUMNAR ADDITION

MATERIALS NEEDED

• pencils
• objects for counting (ones)

Concept:

add three numbers of one digit

Teacher Goal:

To teach the children to add three numbers of one digit.

Teaching Pages 5 and 6:

Tell the children to use the objects for counting to make three sets of numbers—a set of *3*, a set of *2*, and a set of *1*. Have them add the sets of *3* and *2* together and ask how many they have. (5) Have them add the set of *1* to the new set and ask how many they have. (6)

Turn to page 5. Tell the children that today they will learn to add three numbers together. Point to the first example. Explain that they add the *3* and *2* together and then the *5* and *1*.

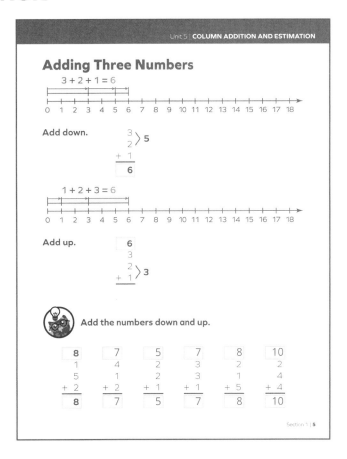

Compare this to the sets that they have just worked with. Have the children point to the number *3* on the number line and count plus *2* and plus *1*. Point to the first example again. Tell the children that this is called *adding down*. Point to the second example. Explain to the children that they can find the same answer by *adding up*. Go through the same process using objects for counting but add sets of *1 + 2* (3) and then add a set of *3*. (6) Have the children point to the numbers in the problem and *add up*. Then have them add on the second number line. Ask them if they have the same answer whether they *add down* or *add up*. Point to the next set of instructions. Tell the children that they should *add down* first and put the answer in the box below the problem. Then they should add up and put the answer in the box above the problem. Go through each problem in this manner. *One* plus *five* is *six*; *six* plus *two* is *eight*. *Two* plus *five* is *seven*; *seven* plus *one* is *eight*.

Turn to page 6. Have the children complete this page *adding down* and *adding up*. Some children may need to jot down their subtotals at first. Watch them carefully to be sure they are not just rewriting the total from the box below to the box above. If necessary, have them say each problem orally as they work it.

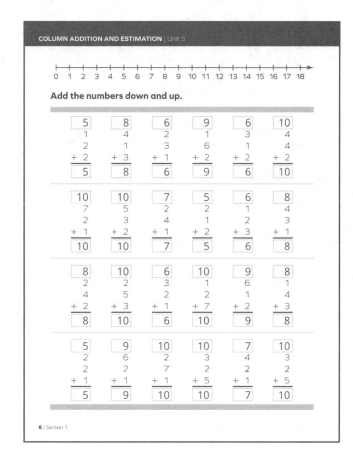

PAGE 7: COLUMNAR ADDITION

MATERIALS NEEDED

• pencils
• objects for counting
• addition fact cards

Concept:

add three numbers of one digit

Teacher Goal:

To teach the children to add three numbers of one digit.

Teaching Page 7:

Turn to page 7. Tell the children that these problems are just like the problems that they worked on pages 5 and 6. Because the numbers are larger on this page, lines have been provided for subtotals. The lines on the left are provided for the subtotal when *adding down* and the lines to the right are for subtotals when *adding up*. Work the first problem with the students to be sure they understand the process. Remind them that the numbers on the left and right will be different. It is important that the students complete this part of the exercise so that the teacher can be sure the students are going through the process of *adding down* and *adding up*. Students should use the number line or objects for counting if they are having difficulty finding the correct answers.

Review addition facts using fact cards for *5's*, *6's*, and *7's* and any additional facts with which the students are having difficulty.

The students should prepare for the Self Test. Ask the students to look over and read the Self Test but they should not write the answers to any questions. After looking over the Self Test the students should go to the beginning of the unit and reread the text and review the answers to the activities up to the Self Test.

The students are to complete the Self Test the next school day. This should be done under regular test conditions without allowing the students to look back. A good idea is to clip the pages together before the test.

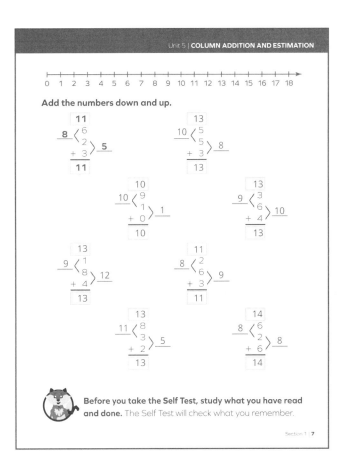

PAGE 8: SELF TEST 1

MATERIALS NEEDED

• pencils

Concept:

columnar addition, number sentences

Teacher Goal:

To teach the children to learn to check their progress periodically.

Teaching Page 8:

Turn to page 8. Read the directions to the children. Have the children repeat them after you while running their fingers under the sentence being read. Be sure the children understand what they are to do. They may use the number line to help them solve the addition problems. You may repeat the directions but give no other help.

Do not have the children check their own work. Check it as soon as you can, and go over it with each child. Show him where he did well and where he needs extra help.

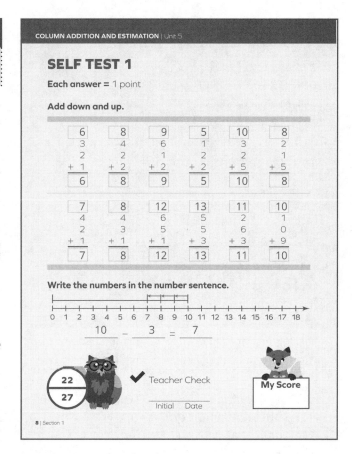

2. NUMBER ORDER

PAGES 9 & 10: COUNT BY 2'S AND 10'S

MATERIALS NEEDED

- pencils
- chart of numbers
- pennies, dimes

Concept:

count numbers by 2's and 10's

Teacher Goal:

To teach the children to skip count by 2's and 10's.

Teaching Pages 9 and 10:

Review skip counting by *2's* with the students using their chart of numbers. Give the students a set of *ten* pennies and have them count them individually. Then have them divide them into sets of *twos* and skip count to *ten*.

Turn to page 9 and read the directions with the students. Tell them to point to the pennies and count to *ten*. Have them complete the page by filling in the missing numbers. When the page is finished, have them read the numbers aloud.

Show the students a dime and have them count how many pennies equal one dime. Review skip counting by *10's* with the students using their chart of numbers. Point out to them that skip counting by *10* is the same as counting dimes. Have them skip count to *100* using dimes.

Turn to page 10 and read the directions with the students. Have them point to the dimes and count to *one hundred*. Emphasize that *ten* dimes are equal to *one hundred* pennies. Tell the students to complete the page by filling in the missing numbers. When the page is finished, have them read the numbers aloud.

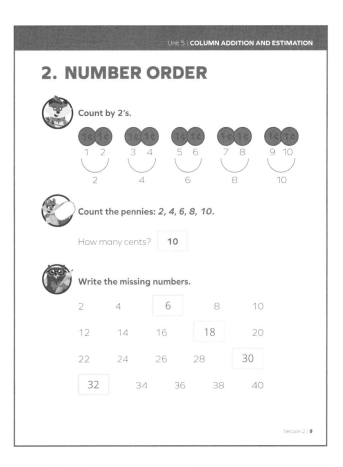

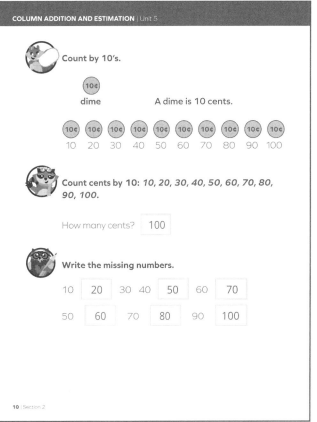

PAGE 11: NUMBER ORDER TO 100

MATERIALS NEEDED

- pencils
- chart of numbers

Concepts:

number order, even and odd

Teacher Goals:

To teach the children to find the missing number, to identify even and odd numbers, and to arrange numbers in number order.

Teaching Page 11:

Have the children skip count by *2's* to *twenty* using their chart of numbers. Ask them if they remember *even* and *odd* numbers. Remind them that the numbers that they say when they count by *2* are the *even* numbers and the ones they skip are the *odd* numbers. Have the children point to the *1* on their chart of numbers and continue through *20* saying "*odd, even, odd, even.*" Point to numbers at random on the number chart and ask if it is an *odd* or *even* number.

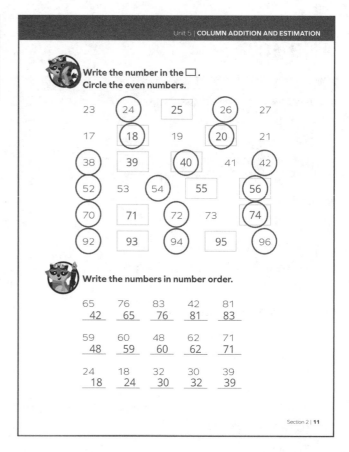

Turn to page 11. Read the directions aloud with the students. Emphasize that there are two sets of directions to the first exercise. After reading the directions to the second exercise, tell the students to point to the number in the *tens'* place and the number in the *ones'* place. Remind them to look at the number in the *tens'* place first when they are putting numbers in number order. Have them complete the page independently.

PAGE 12: NUMBER ORDER TO 100

MATERIALS NEEDED

- pencils
- chart of numbers
- addition fact cards for 6's and 7's
- number symbol cards

Concepts:

number order, even and odd

Teacher Goals:

To teach the children to write the number before and after, to review addition facts, to arrange numbers in number order, and to identify even and odd numbers.

Teaching Page 12:

Turn to page 12 and read the directions aloud with the students. Point out to them that the second exercise has three steps for them to complete.

Monitor the students work to be sure they complete each step of the assignment. Play a game of concentration with the students using the fact cards for *6's* and *7's* and the number symbol cards.

PAGE 13: NUMBER ORDER TO 100

MATERIALS NEEDED

• pencils
• new addition fact cards for 8 + 3, 8 + 4, 8 + 5, 8 + 6, 8 + 7, 8 + 8, and 8 + 9
• number symbol card for 17

Concepts:

greater than, less than, equal, not equal

Teacher Goals:

To teach the children to find the number that is greater than or less than another number and to write = or ≠ to complete number sentences.

Teaching Page 13:

Turn to page 13 and read the directions for the first two exercises with the students. Have the students point to the numbers in the *tens'* place and the numbers in the *ones'* place. Remind them that the number in the *tens'* place is the first number to look at to decide *greater than* or *less than*. Read the directions to the last exercise and then write the symbols for *equal* (=) and *not equal* (≠) on the board. Review the meaning of the signs and give the students several examples. The students should be able to complete the page independently. When the page is finished, use the objects for counting to introduce the addition fact cards for *8's*. Have the students go through the fact cards once.

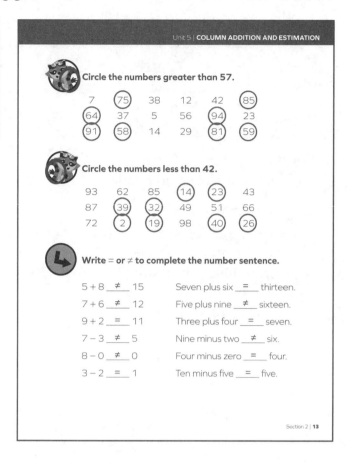

PAGE 14: PLACE VALUE

MATERIALS NEEDED

- pencils
- objects for counting (ones and tens)
- addition fact cards for 8's
- number symbol cards

Concept:

place value for 1's and 10's

Teacher Goal:

To teach the children to identify place value for 1's and 10's.

Teaching Page 14:

Write several numbers on the board and have the children illustrate the numbers using the objects for counting.

For example:

42 = *4* tens and *2* ones. Ask the children what *4* tens is equal to. (40)

Turn to page 14 and read the directions with the students. Point to the blocks that represent *10* and *1*, and review the illustration below it.

Point to *30 + 4* and ask the children, "How many tens in 30?" (3) Have the students write the *3* in the tens' place and *4* in the ones' place:

30 + 4 = 34.

Complete the exercise.

Point to *65* and ask the students, "How many tens and how many ones?" (6 tens and 5 ones) Ask the value of *6* tens and *5* ones. (60 and 5)

65 = 60 + 5.

Complete the exercise.

Read the next set of directions. Tell the children that it is easier to solve three-number problems if they are written vertically (up and down). Tell them to rewrite the problems in the boxes provided and then solve the problems. Have them *add down* and *add up*. Explain to them if they have the same answer, they know their answer is correct. This is called *checking* the answer.

When the page is finished, have the children illustrate the facts for *8's* using the objects for counting. Have them match the fact cards for *8's* with the number symbol cards.

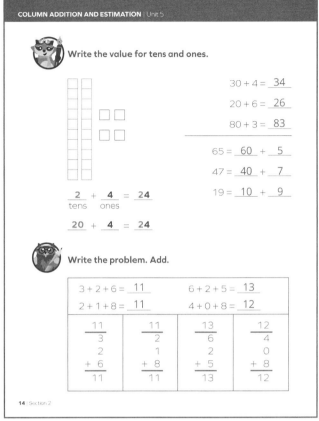

PAGE 15: REVIEWING FACTS

MATERIALS NEEDED

• pencils

Concept:

addition and subtraction facts

Teacher Goal:

To teach the children to review facts for addition and subtraction.

Teaching Page 15:

Turn to page 15 and read the first set of directions with the students. Explain to them that they should add corresponding numbers and put the answer in the box. (2 + 4 = 6) Read the next set of directions. Have the children complete this part of the page independently. Tell them that you are going to dictate four subtraction problems to them.

They should write the problems vertically and in number symbols in the boxes provided. They must also write the answer.

Dictate:

$$9 - 6 \qquad 7 - 3 \qquad 4 - 0 \qquad 8 - 2$$

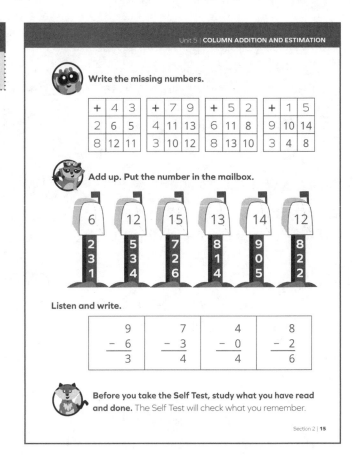

The students should prepare for the Self Test. Ask the students to look over and read the Self Test but they should not write the answers to any questions. After looking over the Self Test the students should go to the beginning of the unit and reread the text and review the answers to the activities up to the Self Test.

The students are to complete the Self Test the next school day. This should be done under regular test conditions without allowing the students to look back. A good idea is to clip the pages together before the test.

PAGE 16: SELF TEST 2

MATERIALS NEEDED

• pencils

Concepts:

number order to 100, equal and not equal, place value

Teacher Goal:

To teach the children to learn to check their progress periodically.

Teaching Page 16:

Turn to page 16. Read the directions to the children. Have the children repeat them after you while running their fingers under the sentence being read. Be sure the children understand what they are to do. You may repeat the directions but give no other help.

Do not have the children check their own work. Check it as soon as you can, and go over it with each child. Show him where he did well and where he needs extra help.

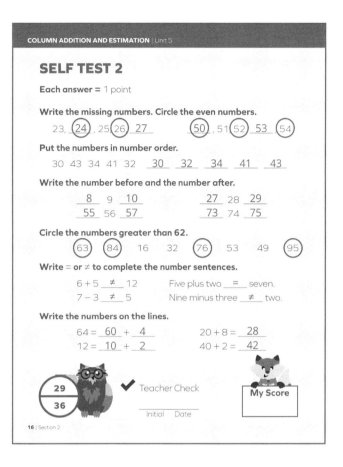

3. FRACTIONS

PAGE 17: FRACTIONS

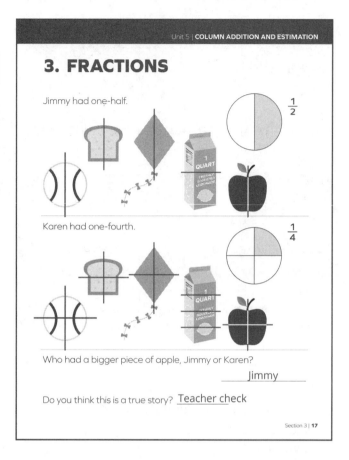

MATERIALS NEEDED

- pencils
- cardboard
- scissors
- 12-inch ruler
- paper
- addition fact cards for 7's and 8's

Concept:

one-half and one-fourth

Teacher Goal:

To teach the children to learn about fractions that show one-half and one-fourth.

Teaching Page 17:

Cut a large circle from a piece of cardboard. Tell the children to use the ruler to divide the circle into *two* parts that are the same size. Let the students cut the circle on the line that they have drawn. Tell them to draw a line on a piece of paper and write the number *2* below the line to represent the number of parts they divided the circle into. Have them hold up *one* of the parts and then write the number *1* above the line. Ask them if they remember what this number is called. (fraction) Have them say the fraction aloud. (one-half)

Cut out another circle. This time have the students use the ruler to divide the circle into *four* parts that are the same size. (It may take them some time to decide how to do this.) Let them cut the circle with the scissors. Tell them to draw a line on paper and write the number *4* below the line to represent the number of parts they divided the circle into. Tell them to hold up *one* part and write the number *1* above the line. Have them say the fraction aloud. (one-fourth) Mix the pieces of cardboard and tell the students to put them back together to form *two* circles.

Turn to page 17. Tell the children to point to the circle that represents *one-half* and the circle that represents *one-fourth*. Have them find the fractions and read the directions that go with them. Ask the children to identify the illustrations on the paper and tell them you are going to read a story.

Jimmy and Karen were brother and sister. One morning Jimmy said to Karen, "I had a funny dream last night. I dreamed it was Saturday morning and I woke up to go out to play. But everything I used was divided in one-half. I had one-half a slice of bread for toast; there was one-half a carton of milk. When I went outside I only had one-half a ball to play with, and there was only one-half a kite to fly. I was so unhappy that I thought I needed a treat. I opened the refrigerator and—oh, no—there it was, one-half an apple!"

Tell the children to draw a line through each one of the pictures to illustrate Jimmy's dream. Then continue by telling the children that Karen told Jimmy that she also had a dream. Read the story over again but substitute *one-fourth* for *one-half*. Have the children draw lines through the second group of pictures to illustrate Karen's dream. Then have them answer the questions at the bottom of the page. When finished, review addition facts for *7's* and *8's*.

PAGE 18: FRACTIONS

MATERIALS NEEDED

- pencils
- objects for counting
- paper

Concept:

one-half and one-fourth

Teacher Goal:

To teach the children to learn about fractions that show one-half and one-fourth.

Teaching Page 18:

Give the children a set of *eight* objects for counting. Ask the children to divide the set into *two* equal parts (4 each part). Tell them to draw a line on a piece of paper and write a *2* below the line to represent the number of parts they divided the set into. Have them hold up the *four* objects. Tell them that the *four* objects represent the part of the set that they are talking about. Have the children write a *1* above the line to repre-

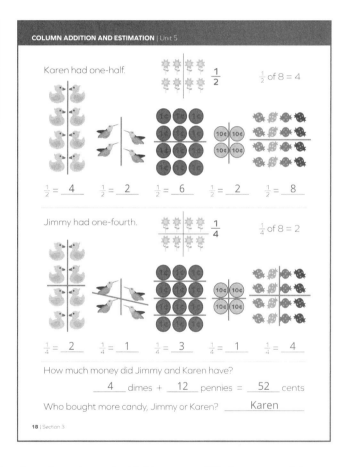

sent one of the *two* parts. Tell them to read the fraction. (one-half) Now say, "If there are *8* in *one* whole set, how many are there in *one-half* a set?" (4) Use the same set of objects and ask the children to divide it into *four* equal parts (2 each part). Tell them to draw a line on paper and write a *4* below the line to represent the number of parts they divided the set into. Ask them to hold up the *two* objects and write a *1* above the line to represent the part of the set that they are talking about. Tell them to read the fraction. (one-fourth) Now say, "If there are *8* in *one* set, how many are there in *one-fourth* a set?" (2)

Turn to page 18 and tell the children to point to the flowers that represent *one-half* and the flowers that represent *one-fourth*. Have the children find the fractions and read the directions that go with them. Ask them to identify the illustrations on the paper and tell them you are going to read a story.

Jimmy and Karen went to the store together. They hurried home to tell their mother about their adventure.

Karen said, "I walked through the park and fed *one-half* of *eight* ducks, and *one-half* of *four* birds. I spent *one-half* of my *twelve* pennies and *one-half* of my *four* dimes. I bought *one-half* of *sixteen* pieces of candy."

Jimmy said, "I walked through the park and fed *one-fourth* of *eight* ducks, and *one-fourth* of *four* birds. I spent *one-fourth* of my *twelve* pennies and *one-fourth* of my *four* dimes. I bought *one-fourth* of *sixteen* pieces of candy."

Tell the children to draw lines through the illustrations to represent Karen's story and Jimmy's story. Then have them answer the questions.

PAGES 19 & 20: CLOCKS

MATERIALS NEEDED

- pencils
- clock from LIFEPAC 103, page 26
- addition fact cards for 7's and 8's
- objects for counting

Concept:

telling time to the hour and half-hour

Teacher Goal:

To teach the children to tell time to the hour and half-hour.

Teaching Pages 19 and 20:

Use the clock the students made to review time to the hour.

Turn to page 19. Read the directions on the page with the students. Have the students illustrate *4:30* on their clocks and have them say *"four thirty"* aloud.

Turn to page 20. Tell the students to illustrate each one of the clocks on the page using their own clock as they say the time aloud. Read the directions and have the students complete the page independently. Have the children illustrate addition facts for *7's* and *8's* using objects for counting.

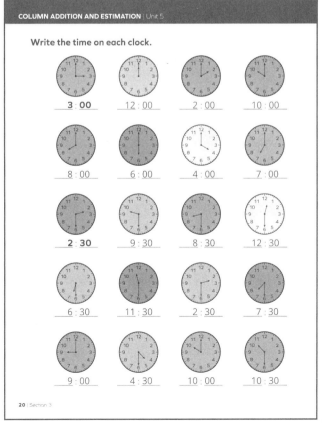

PAGES 21 & 22: CLOCKS, COUNT TO 12

MATERIALS NEEDED

- pencils
- scissors
- large sheet of paper
- paste/glue

Concepts:

clocks, one dozen

Teacher Goals:

To teach the children to learn that clocks count to 12, to learn that there are 12 in a dozen, and to tell time to the hour and half-hour.

Teaching Pages 21 and 22:

Turn to page 21. Read the first set of directions and have the children write the numbers on the clock. Introduce the word *dozen* to the children and talk about where they have used the word *dozen*. Point to the second set of directions and have the students count how many in a *dozen*.

Turn to page 22. Tell the students to write the time on each clock and say the time aloud. Let them cut out the clocks by cutting along the lines that separate the clocks. When they have the clocks cut out in small rectangles tell them to arrange the clocks in number order and paste to a large sheet of paper. (Not all times are represented.) Be sure, for example, that they understand that *1:30* is between *1:00* and *2:00*.

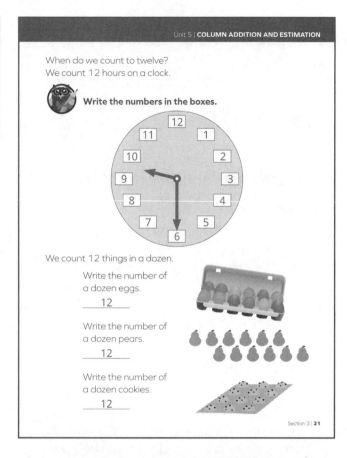

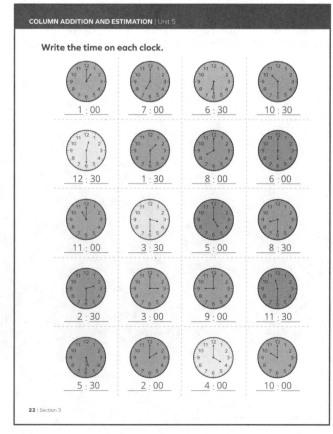

PAGE 23: CLOCKS

MATERIALS NEEDED

- pencils
- crayons
- addition fact cards for 5's and 6's

Concept:

telling time to the hour and half-hour

Teacher Goal:

To teach the children to tell time to the hour and half-hour.

Teaching Page 23:

Turn to page 23. Read the rhyme at the top of the page with the children. Ask them to find the bunny, the bunny's home, the owl, the grasshopper and the flowers. Have them answer the questions at the bottom of the page and color the picture.

Review the addition facts for *5's* and *6's*.

The students should prepare for the Self Test. Ask the students to look over and read the Self Test but they should not write the answers to any questions. After looking over the Self Test the students should go to the beginning of the unit and reread the text and review the answers to the activities up to the Self Test.

The students are to complete the Self Test the next school day. This should be done under regular test conditions without allowing the students to look back. A good idea is to clip the pages together before the test.

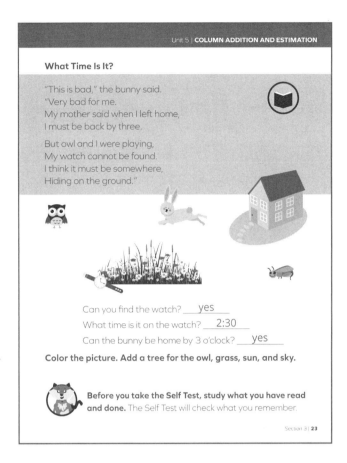

Unit 5 | **COLUMN ADDITION AND ESTIMATION**

What Time Is It?

"This is bad," the bunny said.
"Very bad for me.
My mother said when I left home,
I must be back by three.

But owl and I were playing,
My watch cannot be found.
I think it must be somewhere,
Hiding on the ground."

Can you find the watch? ___yes___
What time is it on the watch? ___2:30___
Can the bunny be home by 3 o'clock? ___yes___

Color the picture. Add a tree for the owl, grass, sun, and sky.

Before you take the Self Test, study what you have read and done. The Self Test will check what you remember.

Section 3 | **23**

PAGE 24: SELF TEST 3

MATERIALS NEEDED

• pencils

Concepts:

fractions, telling time, dozen

Teacher Goal:

To teach the children to learn to check their progress periodically.

Teaching Page 24:

Turn to page 24. Read the directions to the children. Have the children repeat them after you while running their fingers under the sentence being read. Be sure the children understand what they are to do. You may repeat the directions but give no other help.

Do not have the children check their own work. Check it as soon as you can, and go over it with each child. Show him where he did well and where he needs extra help.

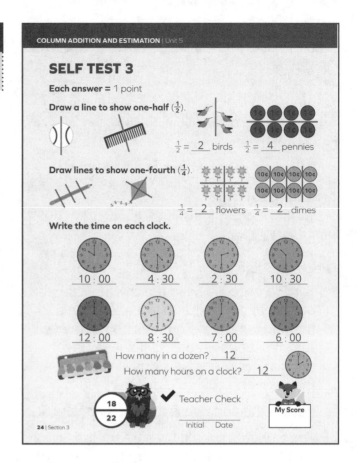

4.   ORDINAL NUMBERS

PAGE 25: ORDINAL NUMBERS

MATERIALS NEEDED

- pencils
- objects for counting
- subtraction fact cards

Concept:

ordinal (order) numbers

Teacher Goal:

To teach the children to show order using ordinal numbers.

Teaching Page 25:

Place *ten* objects for counting in front of the students and point to them saying this is the *first* object, this is the *second* object, and so on through the *tenth* object. Quiz the students asking them at random to point to the *fourth* object, the *ninth* object, and so on.

Turn to page 25 and read the directions. Go through the first list of *ordinal* words to be sure the students recognize each one. Allow the students to complete the page independently. Review the subtraction facts *0* through *9*.

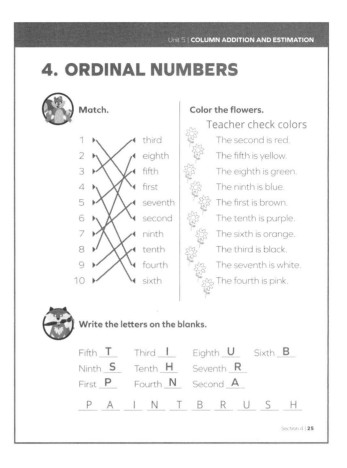

PAGE 26: SUBTRACTION

MATERIALS NEEDED

- pencils
- empty can or plastic glass
- ten beans or similar objects

Concept:

reviewing subtraction to 10

Teacher Goal:

To teach the children to subtract to 10 using number words and number symbols.

Teaching Page 26:

Turn to page 26. Explain to the children that they are going to play a game called *Spill the Beans*.

Have them point to the illustrations and count the number of glasses on the page. Tell them that they should put the beans in their empty can or plastic glass and then knock it over gently so that some but not all of the beans fall out. Have the children count the number that spilled out, write the number in the number sentence, and then write the answer to the sentence. Next, they should count the number of beans left in the can to see if their count is the same as their answer to the problem. Tell them to do this *ten* times.

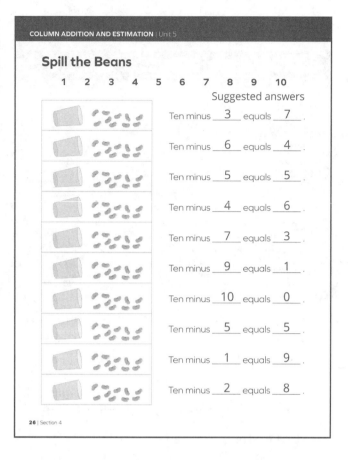

PAGE 27: CHARTS

MATERIALS NEEDED

- pencils
- crayons

Concept:

making a chart

Teacher Goal:

To teach the children to illustrate data by posting it to a chart.

Teaching Page 27:

Turn to page 27. Explain to the children that the picture that they are looking at is called a *chart*. Tell them that they are going to illustrate their game *Spill the Beans* on the *chart*. Read the first question aloud with the students and have them write the answer. Read the next sentence and have them write the numbers *1* to *10* on the lines along the side of the chart. Read the next question. Tell them write the number of beans from *1* to *10* on the lines at the bottom of the chart. Now have them use their crayons to color in how many beans were left in the can each time they *spilled the beans*. They may use the same color crayon or different colors.

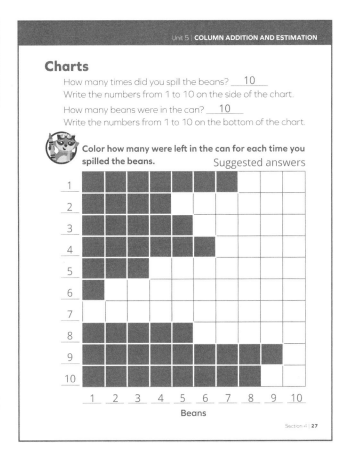

213

PAGE 28: COUNTING MONEY

MATERIALS NEEDED

- pencils
- pennies and dimes
- addition fact cards for 5's through 6's
- number symbol cards

Concept:

pennies and dimes

Teacher Goal:

To teach the children to show how many dimes and how many pennies.

Teaching Page 28:

Review pennies and dimes. Remind the children that the pennies are like the *ones'* strips for counting and the dimes are like the *tens'* strips for counting. Write a number on the board and have the children say how many dimes and how many pennies. Do this several times.

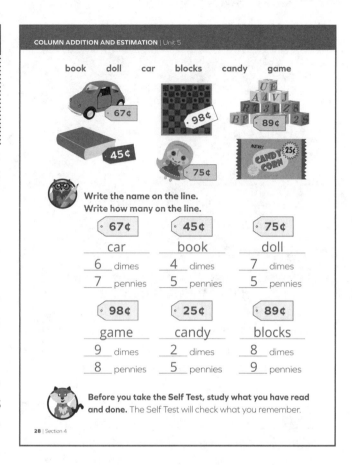

Turn to page 28. Talk about the things that are for sale and how much each item costs. Read the words at the top of the page and the directions with the students. Tell them to write the name of the object that goes with the tag and then write how many dimes and how many pennies.

When finished, have the students match addition fact cards with number symbol cards.

The students should prepare for the Self Test. Ask the students to look over and read the Self Test but they should not write the answers to any questions. After looking over the Self Test the students should go to the beginning of the unit and reread the text and review the answers to the activities up to the Self Test.

The students are to complete the Self Test the next school day. This should be done under regular test conditions without allowing the students to look back. A good idea is to clip the pages together before the test.

PAGE 29: SELF TEST 4

MATERIALS NEEDED

• pencils

Concepts:

ordinal numbers, dimes and pennies, charts

Teacher Goal:

To teach the children to learn to check their progress periodically.

Teaching Page 29:

Turn to page 29. Read the directions to the children. Have the children repeat them after you while running their fingers under the sentence being read. Be sure the children understand what they are to do. You may repeat the directions but give no other help.

Do not have the children check their own work. Check it as soon as you can, and go over it with each child. Show him where he did well and where he needs extra help.

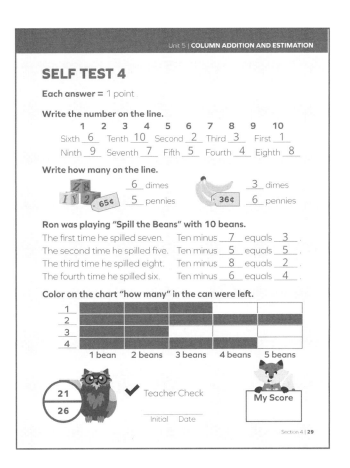

SELF TEST 4

Each answer = 1 point

Write the number on the line.

1   2   3   4   5   6   7   8   9   10

Sixth _6_   Tenth _10_   Second _2_   Third _3_   First _1_

Ninth _9_   Seventh _7_   Fifth _5_   Fourth _4_   Eighth _8_

Write how many on the line.

_6_ dimes   _3_ dimes

65¢ _5_ pennies   36¢ _6_ pennies

Ron was playing "Spill the Beans" with 10 beans.

The first time he spilled seven.   Ten minus _7_ equals _3_ .

The second time he spilled five.   Ten minus _5_ equals _5_ .

The third time he spilled eight.   Ten minus _8_ equals _2_ .

The fourth time he spilled six.   Ten minus _6_ equals _4_ .

Color on the chart "how many" in the can were left.

1					
2					
3					
4					
	1 bean	2 beans	3 beans	4 beans	5 beans

21
26

✔ Teacher Check

Initial   Date

My Score

Section 4 | **29**

215

5. NUMBER LINE

PAGE 30: NUMBER WORDS

MATERIALS NEEDED

- pencils
- addition fact cards for 6's, 7's, and 8's

Concept:

number words to twenty

Teacher Goal:

To teach the children to read and write number words to twenty.

Teaching Page 30:

Turn to page 30. Read the directions with the students. Read the number words aloud with them several times until they say them easily. Have them look for clues to help them recognize the words (*four* in *fourteen*, *"fif"* for *five* in *fifteen*). Let them complete the page independently.

Review fact cards for *6's, 7's* and *8's*.

COLUMN ADDITION AND ESTIMATION | Unit 5

5. NUMBER LINE

 Write the number on the line.

sixteen	16	fourteen	14
twenty	20	seventeen	17
eleven	11	thirteen	13
twelve	12	eighteen	18
nineteen	19	fifteen	15

Write the word on the line.

11	eleven	16	sixteen
12	twelve	17	seventeen
13	thirteen	18	eighteen
14	fourteen	19	nineteen
15	fifteen	20	twenty

Write the number symbol or number word before and after.

15 16 17   eleven twelve thirteen
fifteen sixteen seventeen   11 12 13
18 19 20   thirteen fourteen fifteen
eighteen nineteen twenty   13 14 15

PAGE 31: NUMBER LINE

MATERIALS NEEDED

• pencils

Concept:

add and subtract on a number line

Teacher Goal:

To teach the children to add and subtract on a number line.

Teaching Page 31:

Turn to page 31 and point to the number line at the top of the page.

Read the first set of directions.

Tell the children to put their pencils on the number line at *6* and add *4.*

Tell them to write on the line where they are now on the number line. (10)

Have them subtract *3* and write where they are now on the line. (7)

Have them add *5* and write where they are now. (12)

Have them write the answer to, "Where did you end?" (12)

Have them complete the page in this manner. Their answers should be in number words for the second half of the page.

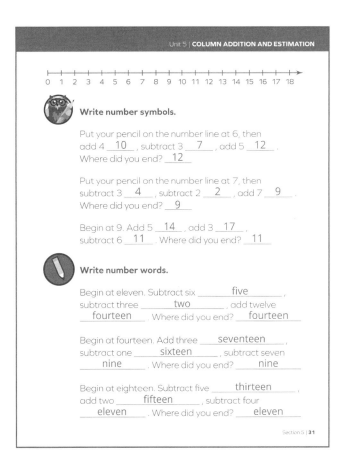

PAGE 32: PATTERNS AND ESTIMATION

MATERIALS NEEDED

• pencils

Concept:

patterns and estimation

Teacher Goals:

To teach the children to estimate answers and to tell what comes next.

Teaching Page 32:

Talk to the children about estimation and patterns. Tell them that often we can predict things by knowing certain information. Give the children some illustrations.

For example: "Since school started, we have always started our math class at nine o'clock. Tomorrow we will start our math class at nine o'clock." Talk about weather patterns, time patterns, and behavior patterns.

Turn to page 32. Explain to the children that the answers they will write are based on patterns.

Allow those children who are able to read the stories and write the answers independently. Other students many need assistance in reading; however, allow them to answer the question as independently as possible and to write their own answers.

COLUMN ADDITION AND ESTIMATION | Unit 5

Patterns and Estimation

Read the story and write the answer.

 Ben caught 1 fish on Monday. He caught 2 fish on Tuesday and 3 fish on Wednesday. How many fish do you think he caught on Thursday? ___4___

 Pat's brother walked 2 blocks the first day, 4 blocks the second day, and 6 blocks the third day. How many blocks do you think he walked the fourth day? ___8___

 Betty was playing "Spill the Beans." The first time she spilled 2, the second time she spilled 3, the third time she spilled 2, and the fourth time she spilled 3. How many do you think she spilled the fifth time? ___2___ the sixth time? ___3___

Jennie and her family were practicing push-ups. Jennie did 10. Her brother did 20. Her mother did 30. How many push-ups do you think her father did? ___40___

Don was saving pennies. The first week he saved five pennies. The second week he saved six pennies. The third week he saved seven pennies. How many pennies do you think Don saved the fourth week? ___eight___

32 | Section 5

PAGE 33: PATTERNS

MATERIALS NEEDED

- pencils
- new addition fact cards 9 + 2, 9 + 3, 9 + 4, 9 + 5, 9 + 6, 9 + 7, 9 + 8, and 9 + 9
- number symbol card 18
- objects for counting

Concept:

patterns

Teacher Goals:

To teach the children to recognize patterns and to tell what comes next.

Teaching Page 33:

Turn to page 33 and read the directions with the students. Tell them they may illustrate or write the words for their answer. Help the students complete the first exercise and fill in the answer. (cat right-side up) Allow them to continue independently but monitor their work to be sure they are finding the patterns. Go over the papers as a class when all students are finished.

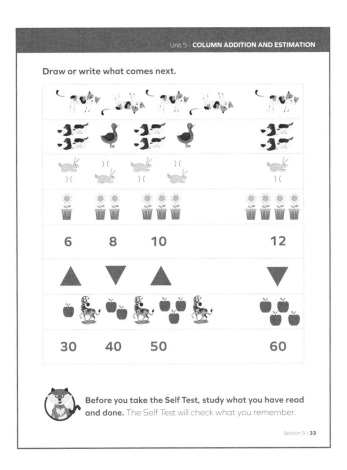

Have the students illustrate the new fact cards for *9's* using objects for counting.

The students should prepare for the Self Test. Ask the students to look over and read the Self Test but they should not write the answers to any questions. After looking over the Self Test the students should go to the beginning of the unit and reread the text and review the answers to the activities up to the Self Test.

The students are to complete the Self Test the next school day. This should be done under regular test conditions without allowing the students to look back. A good idea is to clip the pages together before the test.

PAGE 34: SELF TEST 5

MATERIALS NEEDED

• pencils

Concepts:

number words, patterns

Teacher Goal:

To teach the children to learn to check their progress periodically.

Teaching Page 34:

Turn to page 34. Read the directions to the children. Have the children repeat them after you while running their fingers under the sentence being read. Be sure the children understand what they are to do. You may repeat the directions but give no other help.

Do not have the children check their own work. Check it as soon as you can, and go over it with each child. Show him where he did well and where he needs extra help.

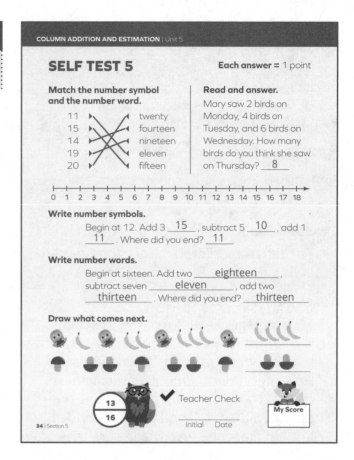

SELF TEST 5 Each answer = 1 point

Match the number symbol and the number word.

11 → twenty
15 → fourteen
14 → nineteen
19 → eleven
20 → fifteen

Read and answer.
Mary saw 2 birds on Monday, 4 birds on Tuesday, and 6 birds on Wednesday. How many birds do you think she saw on Thursday? __8__

0 1 2 3 4 5 6 7 8 9 10 11 12 13 14 15 16 17 18

Write number symbols.
Begin at 12. Add 3 __15__, subtract 5 __10__, add 1 __11__. Where did you end? __11__

Write number words.
Begin at sixteen. Add two __eighteen__, subtract seven __eleven__, add two __thirteen__. Where did you end? __thirteen__

Draw what comes next.

13
16

Teacher Check

Initial   Date

My Score

34 | Section 5

The students should prepare for the LIFEPAC Test. The students should go to the beginning of the unit and reread the text and review the answers to the activities for the entire unit. Ask the students questions to check their understanding of the unit.

The students are to complete the LIFEPAC Test the next school day. This should be done under regular test conditions without allowing the students to look at the unit.

LIFEPAC TEST 105

MATERIALS NEEDED

- pencils
- crayons
- pennies

Concepts:

addition of 3 one-digit numbers, fractions ½ and ¼, place value, time to the half-hour, ordinal numbers, charts, sequence

Teacher Goal:

To teach the children to learn to check their own progress periodically.

Teaching the LIFEPAC Test:

Administer the test in at least two sessions.

Read all of the directions on each page as the children prepare to do it. Be sure that they understand what they are being asked to do.

Give no help except with directions.

Go over each page with the child as soon as possible after you check it so that he can see where he did well and where he needs more work.

Evaluate the tests and review areas where the children have done poorly. Review the pages and activities that stress the concepts tested.

If necessary, when the children have reviewed sufficiently, administer the Alternate LIFEPAC test. Follow the same procedures as used for the LIFEPAC Test.

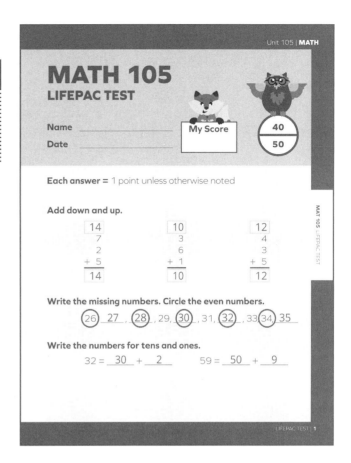

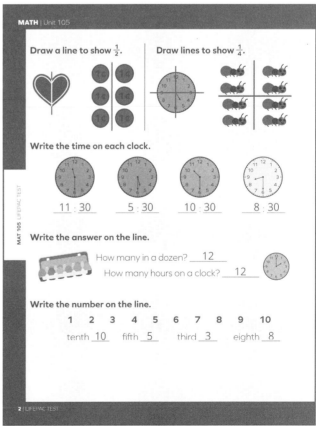

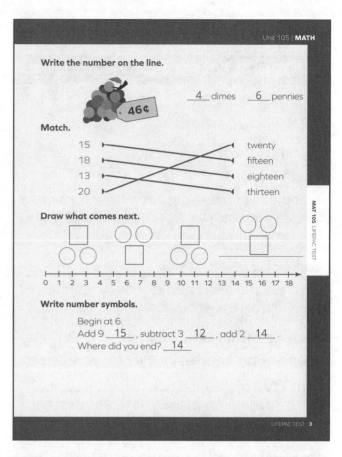

Write the number on the line.

_4_ dimes   _6_ pennies

46¢

Match.

15 ——————→ twenty
18 ——————→ fifteen
13 ——————→ eighteen
20 ——————→ thirteen

Draw what comes next.

0 1 2 3 4 5 6 7 8 9 10 11 12 13 14 15 16 17 18

Write number symbols.

Begin at 6.
Add 9 _15_ , subtract 3 _12_ , add 2 _14_ .
Where did you end? _14_

LIFEPAC TEST | **3**

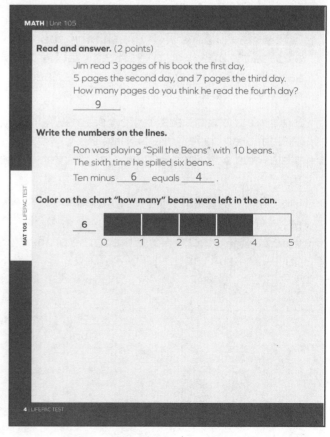

Read and answer. (2 points)

Jim read 3 pages of his book the first day,
5 pages the second day, and 7 pages the third day.
How many pages do you think he read the fourth day?

_9_

Write the numbers on the lines.

Ron was playing "Spill the Beans" with 10 beans.
The sixth time he spilled six beans.

Ten minus _6_ equals _4_ .

Color on the chart "how many" beans were left in the can.

_6_

| 0 | 1 | 2 | 3 | 4 | 5 |

4 | LIFEPAC TEST

ALTERNATE LIFEPAC TEST 105

MATERIALS NEEDED

• pencils
• crayons
• pennies

Concepts:

addition of 3 one-digit numbers, fractions ½ and ¼, place value, time to the half-hour, ordinal numbers, charts, sequence

Teacher Goal:

To teach the children to learn to check their own progress periodically.

Teaching the Alternate LIFEPAC Test:

Administer the test in at least two sessions.

Read all of the directions on each page as the children prepare to do it. Be sure that they understand what they are being asked to do.

Give no help except with directions.

Go over each page with the child as soon as possible after you check it so that he can see where he did well and where he needs more work.

Evaluate the tests and review areas where the children have done poorly. Review the pages and activities that stress the concepts tested.

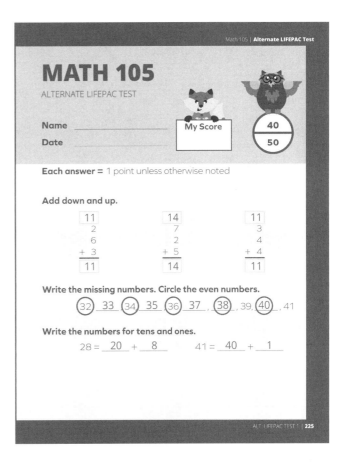

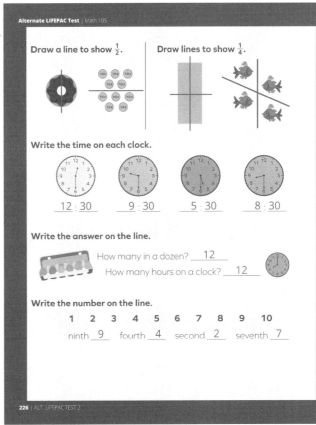

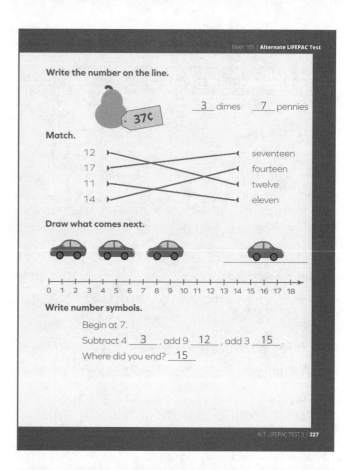

Write the number on the line.

37¢

____3____ dimes   ____7____ pennies

Match.

12   seventeen
17   fourteen
11   twelve
14   eleven

Draw what comes next.

0 1 2 3 4 5 6 7 8 9 10 11 12 13 14 15 16 17 18

Write number symbols.

Begin at 7.
Subtract 4 ___3___ , add 9 ___12___ , add 3 ___15___ .
Where did you end? ___15___

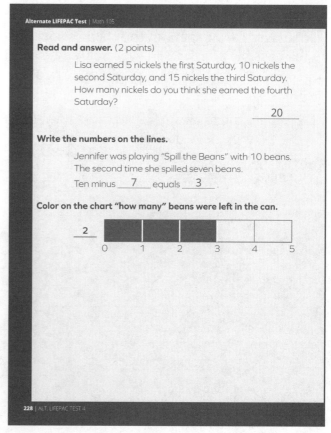

Read and answer. (2 points)

Lisa earned 5 nickels the first Saturday, 10 nickels the second Saturday, and 15 nickels the third Saturday. How many nickels do you think she earned the fourth Saturday?

____20____

Write the numbers on the lines.

Jennifer was playing "Spill the Beans" with 10 beans. The second time she spilled seven beans.
Ten minus ___7___ equals ___3___ .

Color on the chart "how many" beans were left in the can.

___2___

0   1   2   3   4   5

MATH 105

ALTERNATE LIFEPAC TEST

Name _____

Date _____

My Score

40

50

Each answer = 1 point unless otherwise noted

Add down and up.

$$\begin{array}{r} \square \\ 2 \\ 6 \\ + \ 3 \\ \hline \square \end{array} \qquad \begin{array}{r} \square \\ 7 \\ 2 \\ + \ 5 \\ \hline \square \end{array} \qquad \begin{array}{r} \square \\ 3 \\ 4 \\ + \ 4 \\ \hline \square \end{array}$$

Write the missing numbers. Circle the even numbers.

32, _____ , 34, _____ , 36, _____ , _____ , 39, _____ , 41

Write the numbers for tens and ones.

28 = _____ + _____   41 = _____ + _____

Draw a line to show $\frac{1}{2}$.

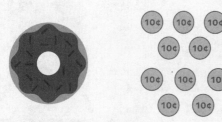

Draw lines to show $\frac{1}{4}$.

Write the time on each clock.

___ : ___   ___ : ___   ___ : ___   ___ : ___

Write the answer on the line.

How many in a dozen? _____

How many hours on a clock? _____

Write the number on the line.

1   2   3   4   5   6   7   8   9   10

ninth ____   fourth ____   second ____   seventh ____

Write the number on the line.

_____ dimes   _____ pennies

Match.

12 ▶   ◀ seventeen

17 ▶   ◀ fourteen

11 ▶   ◀ twelve

14 ▶   ◀ eleven

Draw what comes next.

     _____

```
├──┼──┼──┼──┼──┼──┼──┼──┼──┼──┼──┼──┼──┼──┼──┼──┼──┼──→
0  1  2  3  4  5  6  7  8  9  10 11 12 13 14 15 16 17 18
```

Write number symbols.

Begin at 7.

Subtract 4 _____ , add 9 _____ , add 3 _____ .

Where did you end? _____

Read and answer. (2 points)

Lisa earned 5 nickels the first Saturday, 10 nickels the second Saturday, and 15 nickels the third Saturday. How many nickels do you think she earned the fourth Saturday?

_____

Write the numbers on the lines.

Jennifer was playing "Spill the Beans" with 10 beans. The second time she spilled seven beans.

Ten minus _____ equals _____ .

Color on the chart "how many" beans were left in the can.

2
___

0   1   2   3   4   5

GLOSSARY

Math Terms Glossary

MATH TERMS GLOSSARY

acute angle .... An angle that is less than a right angle or less than 90 degrees.

addend .... A number to be added in an addition problem.

angle .... The distance between two rays or line segments with a common endpoint.

area .... The measurement of a flat surface. $A = l \times w$ (rectangle); $A = \pi r^2$ (circle); $A = \frac{1}{2} b \times h$ (triangle).

associative property .... No matter how numbers are grouped in addition and multiplication, the answer is always the same.

average .... The total of a group divided by the number in the group.

bar graph .... A graph that uses bars to show data.

base (1) .... The bottom part of a geometric figure on which the figure rests.

base (2) .... The number used as a factor in exponential notation.

cancelling .... Simplifying a problem in multiplication or division of fractions within the problem.

cardinal numbers .... Numbers used for counting. 1, 2, 3, 4 ...

Celsius .... Metric unit of measurement for temperature. Freezing, 0° C. Boiling, 100° C.

chart .... An arrangement of data in a logical order.

circle .... A continuous closed line always the same distance from a center point.

circle graph .... A circular graph that always represents the whole of the data.

circumference .... The distance around (perimeter) a circle. $C = 2\pi r$ or $C = \pi d$

common denominator .... Fractions must have the same or common denominator to be added or subtracted.

compass .... An instrument having two hinged legs used for drawing circles, curved lines, and measuring distances.

composite number .... A number that can be divided by 1, by itself, and other numbers.

commutative property .... No matter what order numbers are added or multiplied, the answer is always the same.

congruent .... Figures that have the same size and shape.

cross multiplication .... Multiplying the numerators and denominators of two fractions.

cube .... A solid shape with six square faces.

cylinder .... A round shape with flat ends.

data .... A list of facts from which a conclusion may be drawn.

decimal number .... A fraction with an understood denominator of 10, 100, 1,000 ...

decimal point .... A dot separating the whole number from the fractional part of a decimal number.

degree .... The unit of measurement for angles.

denominator .... The bottom number of a fraction. This number represents the whole.

diameter .... The distance across a circle straight through the center.

difference .... The answer to a subtraction problem.

digit .... Symbols 0, 1, 2, 3, 4, 5 ,6, 7, 8, 9 which when used alone or in combinations represent a value.

dividend .... The number being divided in a division problem.

division bar .... The line that separates the numerator from the denominator of a fraction.

divisor .... The number doing the dividing in a division problem.

endpoints .... Dots that show the beginning and end of a line segment.

equal to .... Has the same value as. (equal, = ; not equal, ≠)

equation .... A number sentence that contains an equal sign.

equilateral triangle .... A triangle whose sides are all equal in length.

equivalent fractions .... Two or more fractions of equal value. To make an equivalent fraction, multiply or divide the numerator and denominator by the same number.

estimate .... To find an approximate answer.

even number .... Any number divisible by two.

expanded form .... Expressing a number by showing the sum of the digits times the place value of each digit.

exponent .... The number that tells how many times a base number is used as a factor.

exponential notation .... Writing a number with a base and its exponent.

face .... The surfaces of a solid figure.

factor(s) .... Numbers which when multiplied together form a product or multiple.

Fahrenheit .... U.S. standard measurement for temperature. Freezing, 32° F. Boiling, 212° F.

fraction .... A number that represents all or part of a whole.

fraction bar .... Also called the division bar.

frequency distribution .... The number of times data falls within a particular classification.

gram .... Metric unit of the measurement of weight.

graph .... A special kind of chart. The most common are bar, line, picture, and circle.

greater than .... Has larger value than. 2 > 1.

greatest common factor .... The largest factor that can be divided into two numbers.

hexagon .... A six-sided polygon.

horizontal .... Level to or parallel to the horizon.

improper fraction .... A fraction that is greater than or equal to 1. The numerator is larger than or equal to the denominator.

input .... Data entered into a calculator (computer).

International Date Line .... The 180th meridian. People who cross the line going west gain a day. People who cross going east lose a day.

intersecting lines .... Lines that cross each other.

invert .... To turn around the positions of the numerator and denominator of a fraction.

isosceles triangle .... A triangle that has two sides of equal length.

least common multiple .... The smallest multiple that two numbers have in common.

less than .... Has smaller value than. 1 < 2.

line .... A continuous set of dots that has no beginning and no end.

line graph .... A graph that shows data by connecting points with lines.

line segment .... The part of a line that has a beginning and an end.

liter .... Metric unit of liquid or dry measurement.

mean .... The same as the average.

median .... The number located exactly in the middle of a list of numbers.

meter .... Metric unit of linear (line) measurement.

minuend .... The number from which another number is being subtracted in a subtraction problem.

mixed number .... A number that combines a whole number and a fraction.

mode .... The number that appears most often in a list of numbers.

multiple .... A multiple of a number is a product of that number.

multiplicand .... The number being multiplied in a multiplication problem.

multiplier .... The number doing the multiplying in a multiplication problem.

negative number .... A number with a value less than zero.

norm .... A standard for a particular group.

number line .... A line with even spaces used to represent certain values.

numeral .... A figure that stands for or represents a number.

numerator .... The top number of a fraction. This number represents the parts being described.

obtuse angle .... An angle greater than a right angle (90 degrees) but less than a straight line (180 degrees).

octagon .... An eight-sided polygon.

odd number .... Any number that cannot be divided evenly by two.

ordered pairs .... Two numbers written in a particular order so that one can be considered the first number and the other the second number.

ordinal numbers .... Numbers that show position. 1st, 2nd, 3rd, 4th ...

output .... The answer to data entered into a calculator (computer).

oval .... A flattened circle; egg-shaped.

parallel lines .... Lines that are always the same distance apart.

pattern .... A set arrangement or design of forms, colors, or numbers.

pentagon .... A five-sided polygon.

percent .... The relationship between a part and a whole. The whole is always 100.

perimeter .... The distance around the outside of a closed figure.

perpendicular lines .... Lines that form right or 90-degree angles.

pictograph .... A graph that uses pictures to represent data.

pi (π) .... Approximately 3.14. Used to solve for the circumference or area of a circle.

place value .... The value of a digit determined by its position in a number.

plane shape .... A flat shape. A plane shape is two-dimensional.

point of intersection .... The one and only point that intersecting lines have in common.

polygon .... A closed plane figure with three or more sides.

positive number .... A number with a value greater than zero.

prediction .... The act of telling something in advance.

prime factorization .... Prime factors of a number expressed in exponential notation.

prime meridian .... The longitudinal meridian (0 degrees) that passes through Greenwich, England.

prime number .... A number divisible by only 1 and itself.

probability .... The study of the likelihood of events.

product .... The answer to a multiplication problem.

proper fraction .... A fraction greater than 0 but less than 1. The numerator is smaller than the denominator.

property of zero .... In addition, any number added to zero will have itself as an answer. In multiplication, any number multiplied by zero will have zero as an answer.

proportion .... An equation stating that two ratios are equal.

protractor .... A semi-circular instrument marked in degrees used to find the measure of an angle.

pyramid .... A solid figure with a polygon as a base and triangular faces that meet at a point.

quadrilateral .... A four-sided polygon.

quotient .... The answer to a division problem.

radius .... The distance from the center of a circle to the edge of a circle. The radius is half of the diameter.

random sample .... A sample in which every member of a large group has an equal chance of being chosen.

ratio .... The relationship of two numbers to each other written 1:2 or 1/2.

ray .... A line with one endpoint.

reciprocal .... The fraction that results from inverting a fraction.

rectangle .... A four-sided polygon with four right angles.

rectangular solid .... A solid figure with six rectangular faces.

reduced fraction .... A fraction equivalent to another fraction that has been written in smaller numbers. This is also called simplifying a fraction or reducing to lowest terms.

remainder .... The amount that remains when a division problem has been completed.

right angle .... An angle that measures 90 degrees.

right triangle .... A triangle with one right angle.

Roman numerals .... The ancient Roman numeral system. (I = 1   V = 5   X = 10   L = 50   C = 100   D = 500   M = 1,000)

scalene triangle .... A triangle with no equal sides.

sequence .... Numbers arranged in a certain pattern.

similar .... Figures that have the same shape but not necessarily the same size.

solid shape .... A shape that takes up space. A solid shape is three-dimensional.

sphere .... A geometric solid in a round shape.

square .... A rectangle with all sides equal.

straight angle .... An angle that measures 180 degrees.

subtrahend .... The number being taken away or subtracted in a subtraction problem.

sum .... The answer to an addition problem.

symmetry .... Shapes with equal halves.

triangle .... A three-sided polygon.

vertex .... The point at which two rays or line segments meet.

vertical .... Straight up and down. Perpendicular to the horizon.

volume .... The measurement of space that a solid figure occupies. $V = l \times w \times h$

whole numbers .... Digits arranged to represent a value equal to or greater than a whole.

METRIC CHART OF PREFIXES

(smallest)	milli-	a unit contains 1,000
	centi-	a unit contains 100
	deci-	a unit contains 10
	unit	unit (meter, liter, gram)
	deca-	contains 10 units
	hecto-	contains 100 units
(largest)	kilo-	contains 1,000 units

ENGLISH SYSTEM OF WEIGHTS AND MEASURES

LENGTH	WEIGHT	DRY MEASURE	LIQUID MEASURE
12 inches = 1 foot	16 ounces = 1 pound	2 cups = 1 pint	16 fl. ounces = 1 pint
3 feet = 1 yard	2,000 lbs. = 1 ton	2 pints = 1 quart	2 cups = 1 pint
36 inches = 1 yard		8 quarts = 1 peck	2 pints = 1 quart
5,280 feet = 1 mile		4 pecks = 1 bushel	4 quarts = 1 gallon
320 rods = 1 mile			

CONVERSION CHART

TO CONVERT	TO	MULTIPLY BY	TO CONVERT	TO	MULTIPLY BY
Linear Measure			**Linear Measure**		
centimeters	inches	.394	inches	centimeters	2.54
meters	yards	1.0936	yards	meters	.914
kilometers	miles	.62	miles	kilometers	1.609
Liquid Measure			**Liquid Measure**		
liters	quarts	1.057	quarts	liters	.946
Dry Measure			**Dry Measure**		
liters	quarts	.908	quarts	liters	1.101
Weight			**Weight**		
grams	ounces	.0353	ounces	grams	28.35
kilograms	pounds	2.2046	pounds	kilograms	.4536